Drinking Water Treatment for Developing Countries
Physical, Chemical and Biological Pollutants

# *Drinking Water Treatment for Developing Countries*
## *Physical, Chemical and Biological Pollutants*

By

**Aniruddha Bhalchandra Pandit**
*Institute of Chemical Technology, Mumbai, India*
*Email: ab.pandit@ictmumbai.edu.in*

and

**Jyoti Kishen Kumar**
*Formerly Institute of Chemical Technology, Mumbai, India*
*Email: jkishenkumar@gmail.com*

ROYAL SOCIETY OF CHEMISTRY

Print ISBN: 978-1-78801-019-1
PDF ISBN: 978-1-78801-293-5
EPUB ISBN: 978-1-78801-762-6

A catalogue record for this book is available from the British Library



The Royal Society of Chemistry is a charity, registered in England and Wales, Number 207890, and a company incorporated in England by Royal Charter (Registered No. RC000524), registered office: Burlington House, Piccadilly, London W1J 0BA, UK, Telephone: +44 (0) 20 7437 8656.

Visit our website at www.rsc.org/books

Printed in the United Kingdom by CPI Group (UK) Ltd, Croydon, CR0 4YY, UK

# Foreword

Water is aptly called 'the elixir of life' and its role in the sustenance of life on earth cannot be over emphasized. High quality drinking water plays a huge role in the survival and health of mankind. Clean and safe drinking water can be assured if the source water is relatively free of physical, chemical and biological contaminants. Therefore, effective treatment strategies that will render water safe for potable use are in great demand. This is a challenge faced by most countries worldwide, but the impact is pronounced in developing nations due to socio-economic constraints on the people living there.

In continuation of the earlier book written by the same authors titled *Drinking Water Disinfection Techniques*, this book has tried to address similar issues, especially in the context of developing countries, since the challenges faced by such countries in translating the known water disinfection techniques to their situation are unique. The book addresses many traditional techniques, which have been practiced for years by the local communities, depending on the available local resources and affordability. The same techniques now appear in the form of engineered knowledge rather than driven by the artisan skill set.

It was nice to see the reference to the book titled *Water Catchers*, which has qualitatively described many such attempts of water harvesting based on need and employing **frugal engineering techniques**, which appear to be the motivation for this book. Needless to say that the treatment methods must be simple, easy to implement and most importantly, cost effective.

In fact, more precisely, for assured success, the methods must meet the **ASSURED** criteria created by me recently, which stand for

**A** (affordable), **S** (scalable), **S** (sustainable), **U** (user friendly), **R** (rapid), **E** (excellent) and **D** (distinctive).

Drinking Water Treatment for Developing Countries: Physical, Chemical and Biological Pollutants
By Aniruddha Bhalchandra Pandit and Jyoti Kishen Kumar

Published by the Royal Society of Chemistry, www.rsc.org

Due to a significant reduction in the ground water level, as a result of excessive draw and a general decline in the quality and quantity of surface-water sources, it is important to look at non-conventional water sources such as sewage water, industrial effluent and harvested rain water as an alternative water source. The authors have addressed this issue quantitatively in terms of an engineered solution, keeping in mind the matrix of "**ASSURED**" as explained above.

The topics covered by Pandit and Jyoti include the now well-established and well-practiced (though on an individual and small community level) technique of **Solar water disinfection** ('**SoDis**') and its hybrid variants, the use of traditional herbal techniques, newer innovative techniques such as the water purifying bicycle, disinfecting hand pump *etc.* This makes the book an interesting read. The quantitative description of these techniques along with the cost comparison depending on the scale of operation, allows the readers, researchers and practitioners to make a judicious choice.

The quantitative interpretation of various case studies validates earlier recommendations. More importantly, it should help the researchers to modify/improve and innovate even further in this area of water management.

I hope the thought and action leadership shown in the book will help society at large on this vital global challenge of 'quality water for all'.

R. A. Mashelkar FRS
National Research Professor
Former President, Global Research Alliance

# Preface

Water treatment has always been an ever-evolving area of study globally and it is particularly important when the treatment is for providing safe drinking water to human beings. Water is required for existence and for the sustenance of normal metabolic activities in humans and other animals, and hence, in addition to being aesthetically appealing, it must be devoid of any pollutants to ensure the safe health of the end users. However, it is common to see that water in most parts of the world is becoming progressively unfit for human consumption. This is largely because of the myriad pollutants that find their way into surface and groundwaters. Industrialization, snowballing environmental pollution, and countless anthropological activities that lead to the contamination of water with sewage and other pollutants are some of the main reasons for the declining quality of drinking water worldwide. This challenge is further magnified in the developing nations as the economic conditions of people living in such regions are generally poor and they cannot employ expensive water treatment techniques that are routinely used in the developed regions globally. Therefore, water treatment methods that are simple, easily accessible and low-priced are the ones that would be acceptable to the rural and developing countries of the world and such techniques have been addressed in this book.

Drinking water treatment was the focus of our previous book, titled *Drinking water Disinfection Techniques*. However, we discussed water disinfection methods that can be used globally and many of those techniques may not be feasible in the milieu of developing nations. Hence, this book was conceptualized keeping the constraints faced by developing countries in mind, which cause specific and yet unique challenges (both social and economic). It is broadly based on the paper published by us in the journal

Drinking Water Treatment for Developing Countries: Physical, Chemical and Biological Pollutants
By Aniruddha Bhalchandra Pandit and Jyoti Kishen Kumar

Published by the Royal Society of Chemistry, www.rsc.org

*Annual Reviews of Chemical and Biomolecular Engineering* in 2015, titled 'Clean Water for Developing Countries', but substantially elaborated with much more quantitative information and the inclusion of several case studies. A nice little book titled *Water Catchers* published by Nimby books, new Delhi, India, listing out inspirational stories, where communities have used and exploited the traditional systems of the collection and storage of water to successfully emerge victorious from desperate situations, was a source of inspiration to complete this book writing exercise.

The basic emphasis of this book is to describe and qualitatively compare various water treatment techniques that can be successfully used in developing countries. These include both the traditional and/or conventional methods and the novel emerging techniques for water treatment at small and large scale, for potable as well as for other domestic activities. An attempt has been made to provide a basis for the selection of one or more of these techniques based on the pollutants to be removed, scale of operation and the cost of the treatment system along with an end use specification of the treated water.

Water conservation is also of paramount importance due to the severe water crunch faced by many nations. Although various ways to save water are being adopted worldwide, there is still scope for further improvisation. One of the highlights of this book is the discussion of possible methods for treating harvested rain water in developing countries. Moreover, various options available for waste water reuse that are suitable in the context of developing nations are also discussed, elaborated and recommended.

The specific case studies highlighted and explained herein have been selected from different parts of the globe to include instances where simple and cost-effective technologies have been deployed to cater to different scales of operation, offering yet another perspective to this topic of discussion. Finally, the book presents the readers with a vast array of information including the latest developments on this topic, covers a discussion on various treatment techniques for the removal of physical, chemical, biological and emerging pollutants based on the water quality guidelines and suggests possible research areas for future work as recommendations, in each chapter towards the end. Since this subject is continuously growing, as in the case of any area of research, it is difficult to cover all the work done and published so far. However, we have tried our best to compile all the relevant work and we hope this book finds its place in the shelves of all those readers who are in some way connected with water treatment and management in developing countries.

Aniruddha B. Pandit
Jyoti Kishen Kumar

# Acknowledgements

A book on a topic of global importance and relevance would require a variety of inputs in various forms from a variety of sources.

Professor Aniruddha B. Pandit acknowledges the assistance, discussion at various stages and invaluable suggestions made by his teachers Professor J. B. Joshi and Professor M. M. Sharma. We are also grateful to Dr R. A. Mashelkar, FRS, National Research Professor and Ex-President, Global Research Alliance for taking some time out of his busy schedule to write a foreword to this book. Prof. Pandit also gratefully acknowledges his students, namely Dr Shankar Kausley, Mr. Chandrakant Holkar, Mr Ketan Desai, Mr Jayesh Mevada, Mr Gaurav Dastane, Dr Dipak Pinjari and a score of others for their contributions in discussions, experimentations, typing and corrections *etc*, without which this book writing exercise would not have been completed. He also acknowledges his colleagues, namely Prof. G. D. Yadav (Vice Chancellor, ICT Mumbai), Dr Parag Gogate, Dr B. D. Kulkarni among others for their valuable course corrections and orientational changes from time to time.

Prof. Pandit also acknowledges the University Grant's commission (Govt. of India), Department of Science and Technology and National academics of India for their assistance in various forms and especially Prof. Taba Kalulu, FTWAS, of Université de Kinshasa, Democratic Republic of Congo for his excellent treatise on hybrid Solar water disinfection.

Prof. Pandit also acknowledges the time sacrifice of his immediate family while completing this manuscript.

Dr Jyoti Kishen Kumar expresses her gratitude to her teacher, guide and mentor, Prof. Pandit, for giving her an opportunity to co-author this book. She acknowledges his valuable guidance, input and advice throughout their writing journey. Dr Jyoti also acknowledges Prof. J. B. Joshi, Prof. M. M. Sharma and Prof. G. D. Yadav (Vice Chancellor, ICT Mumbai) for being a

Drinking Water Treatment for Developing Countries: Physical, Chemical and Biological Pollutants
By Aniruddha Bhalchandra Pandit and Jyoti Kishen Kumar

Published by the Royal Society of Chemistry, www.rsc.org

source of inspiration. She is thankful for the constant motivation from Prof. S. S. Lele. Dr Jyoti acknowledges all the students of Prof. Pandit, especially Dr Shankar Kausley and Mr Ketan Desai, for their contribution in the shaping of this book.

Dr Jyoti heartfully acknowledges the love and support of her family, especially her late Father, who continues to be her source of positive energy in all endeavors. She expresses her gratitude to her mother, husband and in-laws for their encouragement and support. This book would not have been possible without the co-operation and critical comments from her dearest daughters, Ranjini and Rajashree.

# *Dedication*

To our respective families for their continued indulgence in allowing us to complete this book.

Drinking Water Treatment for Developing Countries: Physical, Chemical and Biological Pollutants
By Aniruddha Bhalchandra Pandit and Jyoti Kishen Kumar

Published by the Royal Society of Chemistry, www.rsc.org

# Contents




Drinking Water Treatment for Developing Countries: Physical, Chemical and Biological Pollutants
By Aniruddha Bhalchandra Pandit and Jyoti Kishen Kumar



Published by the Royal Society of Chemistry, www.rsc.org

CHAPTER 1

# *Introduction*

## 1.1 Drinking Water: Availability and Need

Drinking water or potable water is water that is clean, safe and free from any physical, chemical or biological pollutants. Water is a natural resource and is available freely as almost 70% of earth's surface is filled with water. Therefore, it can be said that drinking water availability is a human birthright. It is indeed fascinating to note that the water that we drink today has always been around in some form or another as it is recycled in the atmosphere and in our glass, which holds water. However, it must be noted that only 2.5% of the global water is freshwater and the remaining part accounts for saline water and the oceans. From this 2.5%, only about 1% is reported to be accessible while the remainder is locked in the form of glaciers and snow.

Overall, experts state that only around 0.007% of the earth's water is available for human consumption. This amount is not sufficient to cater to the needs of the ever-increasing population on our planet. As per the statistics reported by the United Nations, the current world population is almost 7.6 billion and the birth rate is 4.3 every second. The population is expected to reach 8.6 billion by 2030, 9.8 billion in 2050 and almost 11.2 billion in the year 2100. Therefore, it can be clearly seen that the existing water resource will not be able to meet the augmenting water demands of the growing global population. It is important to note that freshwater availability is not globally well distributed due to differences in geographies, climate and environmental factors, industrialization and an upsurge in human activities.

Most of the developing countries face a major shortage of fresh water and in some instances, even if water is available, it may not be safe for human consumption and may also entail long hours of commuting to fetch it from the source. Water is very essential to sustain life and the human body is

Drinking Water Treatment for Developing Countries: Physical, Chemical and Biological Pollutants
By Aniruddha Bhalchandra Pandit and Jyoti Kishen Kumar

Published by the Royal Society of Chemistry, www.rsc.org

made up of 70% water. Apart from drinking, water is also needed for basic needs such as bathing, cooking and other domestic chores. According to the United Nations, the water usage rate has increased to more than twice the rate of population growth in the last century. It is very distressing to note that by the year 2025, around 1.8 billion people will be living in water scarce regions and almost two-thirds of the world's population will be living in water strained zones mainly due to uncontrolled and extensive water usage, growth and environmental changes.

## 1.2 Efforts Towards Providing Clean Drinking Water

Global water regulating bodies such as the United Nations, WHO and institutions working on water quality and health such as UNICEF and many other national and local bodies have been performing a herculean task of monitoring water quality, charting guidelines and organizing programs for public awareness and safety. Over the last several decades, the world has witnessed tremendous progress in water quality management and many goals laid down by the WHO and the UN from time to time like the Millennium Development Goals (MDGs) have been successfully met.

It is very heartening to note that in the year 2015, 71% of the global population (2.6 billion) had been using a safely managed drinking water service since 1990. This implies that they used a water source that was located on the premises and that this source was freely available and free from contamination. This is indeed a huge milestone as the proportion of the global population using an improved drinking water source between 1990 and 2015 has increased from 76% to 91%. Moreover, 89% of the global population used at least a basic service, which is an improved drinking water source that is close to the premise and generally requires a round commuting trip of less than 30 mins.

Although this is a positive result, there are still many milestones to be achieved as the remaining population (663 million) still lack even a basic drinking water service and it is reported that globally at least 2 billion people use a drinking water source that may be contaminated with feces. This is very alarming since the consumption of contaminated water can result in many water borne diseases such as cholera, diarrhea, typhoid and dysentery. It is reported that each day nearly 1000 children die due to preventable water and sanitation related diarrheal diseases. Therefore, after the MDGs, the UN drafted the Sustainable Developmental Goals (SDGs) that cover a set of 17 different objectives and are mainly considered a universal call to action to end poverty, and to protect and ensure that all people enjoy peace and prosperity.

These 17 goals built on the successes of the MDGs and include new areas such as climate change, innovation, sustainable consumption *etc.* Interestingly, the goals are interconnected and solving one goal will often involve tackling issues related to the other. Goal 6 is 'clean water and sanitation' and the salient points related to clean water are that by 2030, universal and equitable access to safe and affordable drinking water should be available for all and by 2030, water quality should be improved by reducing pollution, minimizing the release of hazardous chemicals, having a proportion of untreated waste water and substantially increasing recycling and safe reuse globally. Most importantly, one of the main goals is to expand international cooperation and capacity building support to developing countries in water related activities and programmes including water harvesting, desalination, water efficiency, waste water treatment, recycling and reuse technologies.[1]

In this book, an attempt has been made to discuss these technologies with respect to their application in developing countries. Yet another global monitoring body is the WHO and UNICEF joint monitoring Programme for water supply, sanitation and hygiene (JMP), which has produced regular reports on the progress of the work related to goal 6, on drinking water, sanitation and hygiene, since 1990. JMP has developed a global database and was responsible for monitoring the progress of the 2015 MDG. It has now been entrusted with the responsibility of monitoring the progress of the 2030 SDG targets related to drinking water, sanitation and hygiene.

Recently, in 2017, JMP released the first update on the first global baseline estimates for the SDG targets related to drinking water. According to this update, the salient conclusion related to drinking water was that there are many people who still lack access to drinking water, especially in the rural areas. Although there are billions of people who have had access to a basic drinking water facility since 2000, these facilities do not provide safe water and the potential risk of consuming contaminated water is significant. Therefore, the risk of contracting water borne diseases is very high, especially among young children who are prone to diseases such as diarrhea, hepatitis, cholera and typhoid.

Yet another key finding in this report is that there are big gaps in services between the urban and rural areas. For instance, it is reported that two out of three people with safe managed drinking water live in urban areas and of the 161 million people who use untreated surface water (from lakes, rivers or irrigation channels), 150 million live in rural areas. This clearly points out that there is a huge requirement to improve access to safe and clean drinking water in the rural regions, which are mainly located in the developing countries across the world. Some of the developing countries are shown in Figure 1.1.

Thus, availability of safe and clean drinking water is the basic right of every individual and all countries globally have the responsibility to ensure that it is provided to all. This is very important in developing countries as the provision of such a basic requirement and improved services related to drinking water will ensure a bright future for the coming generation. Most importantly, enhanced water supply and better management of water resources can boost the economic growth of a country and contribute significantly to poverty reduction.[2]

**Figure 1.1** Developing Countries (Source: http://www.un.org/en/development/desa/policy/wesp/wesp_current/2014wesp_country_classification.pdf).

## 1.3 Water Pollutants and Health Hazards

One of the most important aspects to be understood and considered in any water treatment program is the quality of the source water. It is a well-established fact that water can have a plethora of contaminants depending on the origin of water and the pollutants that find their way into water bodies from external sources. These two major contributors to the quality of water can therefore introduce a spectrum of substances that can pose a challenge in water treatment processes. For instance, variations in water compositions can be expected based on whether the raw water being treated is surface or ground water and the geographical region as well. On the other hand, due to extensive industrial and human activities over decades, waste water and sewage may find their way into water bodies thereby majorly contributing to the high level of contaminants in fresh and stored water sources.

Although describing every pollutant in detail is beyond the scope of this book, it is necessary to understand some major classes of pollutants and the health risk they cause when consumed by humans. This information is needed to comprehend the importance of water treatment, especially in the developing countries where the affected population is large in number but the economic conditions are not conducive to sophisticated treatment methods. The following sections describe the major types of pollutants, highlighting the significant ones and the diseases that can be caused in humans when water is consumed without adequate treatment.

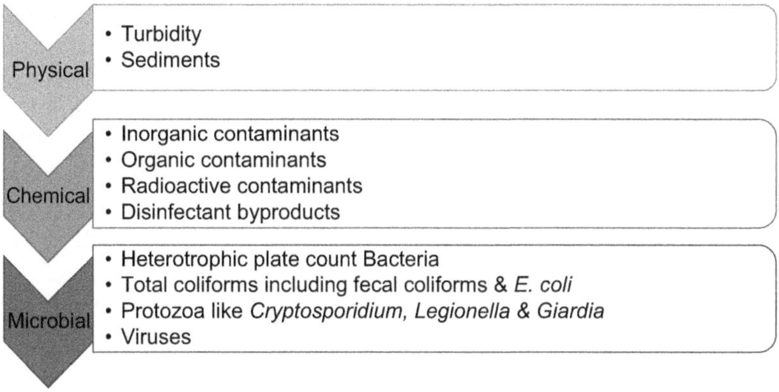


**Figure 1.2** Types of Water Pollutants.

Common water pollutants can be classified into physical, chemical and biological varieties. The major classification is given in Figure 1.2. Water usually contains physical, chemical and microbiological contaminants that pollute water and can pose serious health issues to humans. Physical pollutants include increased levels of turbidity in water, which may be due the presence of sediment or organic contaminants present in it. These factors impart an obnoxious odor and color to water thus rendering it aesthetically unacceptable by humans. These physical contaminants may have originated due to other chemical or microbiological sources that will be discussed in the following sections.

Chemical pollutants can get introduced into water from waste water and industrial effluents. These include a spectrum of inorganic contaminants such as Arsenic, lead, Mercury, Fluoride and Chromium and organic contaminants like Acrylamide, Carbon tetrachloride and vinyl chlorides and many others. Radioactive pollutants such as Uranium, and alpha and beta emitters are also reported to be found in water. For eons, chemical disinfection using chlorine, chlorine dioxide and chloramines has been used to treat water and this has been found to remain in residual amounts in treated water. Although residual disinfectants are required to prevent re-contamination in water distribution systems, high levels of the same can pose a health hazard.

Disinfection byproducts (DBPs) are yet another class of important chemical contaminant that are routinely present in water if chemical disinfection is used to treat drinking water. They are generally formed due to various chemical reactions that occur between the chemical disinfectant and the organic and inorganic polluting matter found in water. Thus bromate, chlorite and trihalomethanes, which are potential carcinogens, are also deleterious compounds that fall under the class of chemical contaminants as DBPs.

Microbial pollutants form a major class of water contaminants, as they can seriously affect human health. Among the microbial contaminants, bacteria, protozoa and viruses constitute the main types that are commonly found in water. Many of these microorganisms are pathogenic in nature and can cause various diseases in humans. Since it is difficult to detect pathogenic microorganisms due to the time it takes for the analysis and the risk involved in handling such specimens, the concept of indicator microorganisms was introduced. *Escherichia coli*, *fecal coliforms* and *fecal streptococci* are a few indicator microorganisms. The presence of indicator organisms in water indicates that it is indeed polluted and is likely to contain pathogenic species as well.

More information on each class of pollutants is included in the following sections. Emphasis on certain key aspects in each class such as their origin, likely route of entry into water, complications and health risks that occur when they are consumed without appropriate treatment is of paramount importance. Some examples of contaminants with a detailed description are also provided as this forms the foundation based on which the importance of safe, effective and economical water treatment can be understood, especially in the context of developing countries where the risks are huge, and the means are poor.

### 1.3.1 Physical Pollutants

The colour and odour of drinking water have a huge role to play and substances contributing to colour and odour in potable water systems are referred to as physical pollutants. This is because humans generally accept and consume drinking water based on its appearance, as water that is aesthetically pure and pleasant is generally approved for potable purposes. However, clear, odorless and colourless water does not imply that it is free from other contaminants such as chemical pollutants and microorganisms. Thus, water that is aesthetically unacceptable can lead to the use of water from sources that are aesthetically more acceptable but possibly unsafe for human health. Therefore, taste and odour of drinking water are of paramount importance and should be acceptable to the consumer.

There are several ways by which the colour, odour and taste of drinking water may change thereby leading to its acceptability issues for the end consumers. Natural organic and inorganic matter, chemical and biological contaminants, synthetic chemicals, corrosion of metal pipes and microbial activity during the storage and distribution of potable water are some of the ways that may contribute to colour and taste changes. However, for all practical purposes, the two main factors are chemically derived pollutants and biological contaminants.

Biological pollutants include Actinomycetes and certain fungi, which are usually present on surface water and they produce a substance called 'goemin' and 2-methyl isobirneol. These substances can lead to taste and odour problems in water. Certain algae and cyanobacteria also produce similar substances and cyanotoxins, which can aggravate the issue and cause problems in the coagulation and filtration processes that are used in water treatment. Iron bacteria thrive in waters that contain high levels of ferrous and manganese salts and they oxidize these substances to produce rust coloured deposits on the walls of tanks and pipes and this in turn can seep into water. Similarly, invertebrate animals such as crustaceans, snails and zebra mussels can also penetrate filters and storage tanks imparting an unpleasant taste and odour to water.

Myriad chemicals can find their way into water resulting in acceptability issues. For instance, when aluminum is present over 0.1 to 0.2 mg $l^{-1}$, it generally results in taste issues. Some chemicals such as chlorine are easily detectable in terms of taste and odour even when present at a concentration below 1 mg $l^{-1}$. The odour threshold of ammonia is around 1.5 mg $l^{-1}$ and for hydrogen sulphide, it is as low as 0.05 to 0.1 mg $l^{-1}$ when consumers can sense the rotten egg odour. Other chemicals such as copper generally get into water from leaching of pipes and above 2 mg $l^{-1}$, copper contributes to colour and taste issues. Colour in water can be due to organic matter such as humic and fluvic acids, iron and other metals, and industrial effluents, and when present above 15 true colour units (TCU), they are detectable visually. Hardness of water due to the presence of calcium and magnesium salts results in precipitation of soap scum and when they are present above 200 mg $l^{-1}$, they can lead to the deposition of scale in pipes and render water with an unpleasant taste.

Total dissolved solid (TDS) when present in low quantities (less than 600 mg $l^{-1}$) generally does not pose a problem. However, when the concentration gets elevated to more than 1000 mg $l^{-1}$, the water becomes unpalatable and results in additional issues such as scaling of pipes and the water distribution system. Yet another important aspect of acceptability of water is its turbidity, which is expressed in terms of nephelometric turbidity units (NTU). Particles such as silt, clay, chemical precipitates and organic impurities contribute to the turbidity of water and when the levels increase to above 4 NTU, it becomes visible and renders the water unacceptable to the end user. It should be less than 0.5 NTU so that the water can be consumed safely. Moreover, augmented levels of turbidity indicate the presence of potential chemical or microbiological contaminants apart from just causing aesthetic issues.[3]

### 1.3.2 Chemical Pollutants

A spectrum of chemical pollutants may be present in drinking water either naturally or introduced due to human and environmental activities. The type of chemical pollutants present is typically dependent on myriad factors such as the water source and storage, geographical region, and possible contamination sites. Most importantly, the type and concentration of the contaminants keep varying and are always in a constant flux. If left untreated, chemical contaminants pose a great threat to humans and therefore they must be identified and subsequently treated appropriately. This becomes of paramount importance in the context of developing countries where the resources and economics of the treatment processes need careful consideration. Over the years, several chemical substances have found their way into raw water such as emerging contaminants like pharmaceutical compounds, fertilizers, pesticides and their derivatives.

Chemical pollutants may be present in the form of inorganic, organic and radioactive contaminants. Often, disinfectants used for water treatment may themselves contribute to the chemical contamination when present in excess quantities. An excess amount of chemical disinfectants is generally used in water treatment schemes to ensure that there is no regrowth of microbes in the distribution system. However, these excess concentrations also react with other organic matter present in the water leading to the formation of disinfection byproducts (DBPs). Most of the chemical pollutants when present at a concentration above the permissible limits laid down by water authorities like the WHO lead to innumerable health hazards in humans. Many of them also contribute to the acceptability issues discussed in the previous section.

Chemical contaminants being innumerable, it is beyond the scope of this book to describe each one in detail. However, some of them along with the health risks associated with their presence in drinking water are described in Table 1.1. Many of them can cause health problems ranging from gastrointestinal disturbances to severe cases of cancer and genotoxicity. Continued efforts by various stakeholders and experts have led to more information on these chemical contaminants and their occurrence in water. Toxicity studies on animal and humans have also thrown light on the potential health hazards, which has helped the national and global regulatory bodies like the WHO to come up with the permissible limits for these chemical pollutants.

New and emerging chemical contaminants in water pose a bigger challenge when coming up with a standard for these pollutants. One such chemical contaminant is methyl tertiary butyl ether (MTBE), which is employed as a fuel additive in the United States. This is primarily done to diminish carbon monoxide and curb ozone depletion triggered by automobile emissions of hydrocarbons. Generally, MTBE is detected at a very low concentration of a few nanograms to micrograms in ground water. However, when present at elevated levels, MTBE can be harmful. Studies indicate that MTBE is a rodent carcinogen at high levels, but it is not genotoxic. The studies carried out to date have not come up with conclusive evidence to point out the possible human carcinogenic nature of MTBE. Therefore, the regulation authorities have not yet come up with the permissible limits for MTBE and efforts on this front are in progress.

### 1.3.3 Microbial Pollutants

One of the greatest risks to human health is the presence of microbial contaminants in drinking water. This is because these pollutants can cause severe illness, which can sometimes be fatal. Therefore, this class of

**Table 1.1** List of some chemical contaminants and their health risks.

| S. no. | Contaminant | Occurrence in water | Health risks | WHO guideline value |
|---|---|---|---|---|
| 1. | Antimony | Its occurrence in water is very low (less than 5 $\mu g\,l^{-1}$) and it generally enters water from leaching of metal pipes and fittings. | International Agency for Research on Cancer (IARC) reports that antimony trioxide is possibly carcinogenic to humans. | 0.02 $mg\,l^{-1}$ (20 $\mu g\,l^{-1}$) |
| 2. | Arsenic | Food and drinking water are the main routes of entry for humans. Its concentration in natural waters is between 1 and 2 $\mu g\,l^{-1}$ and may be higher (up to 12 $mg\,l^{-1}$) in regions having natural sources. | Chronic conditions are dermal lesions such as hyperpigmentation peripheral neuropathy, skin, bladder and lung cancers, peripheral vascular disease. | 0.01 $mg\,l^{-1}$ (10 $\mu g\,l^{-1}$) |
| 3. | Chloramines | In potable water supplies where chloramine is used as a primary disinfectant. It is also used to provide a chlorine residual in the distribution system. Typical concentration is 0.5–2 $mg\,l^{-1}$. | *In vitro* studies show monochloramine to be mutagenic, but it has not been found to be genotoxic *in vivo*. It can cause taste and odour problems. | *Monochloramine:* 3 $mg\,l^{-1}$ (3000 $\mu g\,l^{-1}$) |
| 4. | Nitrate and nitrite | Nitrate levels are generally high in well waters but may be found in other sources as well. High levels of 50 $mg\,l^{-1}$ and above may be found where there are significant sources of contamination. Nitrite levels are normally lower, less than a few milligrams per liter. | Gastrointestinal infections, methaemoglobinaemia, which in infants can give rise to cyanosis, referred to as blue-baby syndrome. Generally, its occurrence is seen in bottle-fed infants. | *Nitrate:* 50 $mg\,l^{-1}$ as nitrate ion<br>*Nitrite:* 3 $mg\,l^{-1}$ as nitrite ion |
| 5. | Fluoride | Added at 0.5 and 1 $mg\,l^{-1}$ to drinking water to combat dental caries. In groundwater, usually 10 $mg\,l^{-1}$ is present with the highest natural level of 2800 $mg\,l^{-1}$. | Effects on skeletal tissues (bones and teeth). Also affects tooth enamel and can cause mild dental fluorosis. | 1.5 $mg\,l^{-1}$ (1500 $\mu g\,l^{-1}$) |

**Table 1.1** (*Continued*)

| S. no. | Contaminant | Occurrence in water | Health risks | WHO guideline value |
|---|---|---|---|---|
| 6. | Uranium | Generally, uranium levels in drinking water are less than 1 $\mu g\,l^{-1}$, however, on some occasions, elevated levels of 700 $\mu g\,l^{-1}$ have also been encountered in some private supplies. | Most common toxic effect seen in humans is Nephritis at high levels of uranium in water. Its carcinogenicity is not yet proven in humans and in animal studies as well. | 0.03 $mg\,l^{-1}$ (30 $\mu g\,l^{-1}$) |
| 7. | Vinyl Chloride | Rarely detected in surface waters, the concentrations measured generally not exceeding 10 $\mu g\,l^{-1}$; much higher concentrations found in groundwater and well water in contaminated areas; concentrations up to 10 $\mu g\,l^{-1}$ detected in drinking water. | Carcinogenic to humans and causes liver cancers, angiosarcomas and hepatocellular carcinoma. Vinyl chloride metabolites are genotoxic. | 0.0003 $mg\,l^{-1}$ (0.3 $\mu g\,l^{-1}$) |
| 8. | Trihalomethanes | THMs are not usually found in raw water but are formed because of chlorination of organic matter. The concentrations are generally below 100 $\mu g\,l^{-1}$ and chloroform is the dominant compound. | IARC has classified chloroform as possibly carcinogenic to humans. It also causes damage to the liver. Bromoform causes genotoxicity and Bromodichloromethane is known to be possibly carcinogenic to humans and increases the risk for spontaneous abortion or stillbirth. | Chloroform: 0.3 $mg\,l^{-1}$ (300 $\mu g\,l^{-1}$) Bromoform: 0.1 $mg\,l^{-1}$ (100 $\mu g\,l^{-1}$) Dibromochloromethane (DBCM): 0.1 $mg\,l^{-1}$ (100 $\mu g\,l^{-1}$) Bromodichloromethane (BDCM): 0.06 $mg\,l^{-1}$ (60 $\mu g\,l^{-1}$) |

pollutants is very important and adequate measures must be taken to avoid drinking microbially contaminated water. Microbes can enter a water body by various ways such as untreated industrial effluents, sewage inflow and other human activities contaminating the water source. Thus, the source water should be kept clean by preventing pollution as far as possible and proper selection of treatment methods and their operation must be carried out to keep these microbes in check. Moreover, the distribution systems should also be monitored, and adequate care should be taken to prevent microbial contamination in the distribution network.

These three aspects, *i.e.* source water, treatment method and distribution systems, are of paramount importance as any breach in any/all of these can result in a major disease outbreak that can even prove to be fatal to humans. This can be associated with the typical nature of the microbial pollutants that makes them different from the physical and chemical pollutants discussed in the preceding sections. Some of these features are due to the fact that many microbes are pathogenic and can cause acute or chronic diseases, they can aggregate with other suspended particles in water and their concentration can vary in time.

The possibility of acquiring an infection from a pathogenic microorganism present in water depends on a plethora of factors such as the virulence of the pathogen, its concentration in water and its ability to invade the host (human body), which in turn also depends on the immune status of the individual. Once the microbes invade the human body, they multiply and cause acute or chronic conditions that could be mild to severe. It is interesting to note that these waterborne pathogens are also capable of multiplying in food and beverages and can thrive in warm water systems like the *Legionella* species (Table 1.2), thereby increasing the possibility of infections in humans.

Microbial pollutants can be of various types such as bacteria, viruses, protozoa and helminths, each of which has myriad members exhibiting varying degrees of pathogenicity. Some of these are described in Table 1.2. It is interesting to note that not all microorganisms infecting humans through the fecal-oral route are transmitted only *via* water. In fact, contamination of food, beverages, unclean hands, and utensils and general poor hygienic conditions of the person and the household are equally responsible for the spread of microbial infections. Moreover, it is important to note that the route of infection cannot be restricted only through drinking contaminated water. Aerosols that contain tiny droplets of water with microbes may also lead to health risks if inhaled by a person, for instance from a shower while bathing. Even cuts, wounds and abrasions of the skin if kept uncovered can be open to invasive microorganisms present in water.

Among the different microbial pollutants that are usually present in contaminated water, bacteria form an important and major class that need careful attention. Most of the bacterial species are ingested while drinking water and they enter the gastrointestinal tract of humans and cause various acute and chronic infections. Subsequently, they get excreted into the feces

**Table 1.2** List of some microbial contaminants and their health risks.

| S. no. | Microbial species | Source and occurrence | Health risks | Prevention of water contamination/ treatment of water |
|---|---|---|---|---|
| 1. | *Campylobacter* spp | Animals, food and water are a significant source. | Acute gastroenteritis, abdominal pain, diarrhea<br>Guillain-Barre syndrome, an acute demyelinating disease of the peripheral nerves.<br>Reactive arthritis and meningitis. | Protection of raw water supplies from waste from humans and animals, adequate treatment and protection of water during distribution. Storage of treated and disinfected water should be protected from bird feces. |
| 2. | *Legionella* spp *E.g. L. pneumophila* | Freshwater, water cooling devices associated with air-conditioning systems, hot water distribution systems and spas, which provide suitable temperatures (25–50 °C). | Legionellosis has two clinical forms called Legionnaires' disease and Pontiac fever. | Monochloramine has been shown to be particularly effective. Wherever possible, water temperatures should be kept outside the range of 25–50 °C to prevent the growth of the organism. |
| 3. | *Enterovirus e.g.*, poliovirus, coxsackievirus B, echovirus | Enteroviruses are excreted in the feces of infected individuals. Most numerous in sewage, water resources and treated drinking water supplies. | Mild febrile illness to myocarditis, meningoencephalitis, poliomyelitis, herpangina, hand-foot and-mouth disease and neonatal multi-organ failure. | Prevention of source water contamination by human waste, followed by adequate treatment and disinfection. Drinking-water supplies should also be protected from contamination during distribution. |

| | | | | |
|---|---|---|---|---|
| 4. | Hepatitis A virus (HAV) | HAV is excreted in fecal material of infected people; fecally contaminated food and water are common sources of the virus. | The virus causes the disease hepatitis A, commonly known as "infectious hepatitis". | Prevention of source water contamination by human waste, followed by adequate treatment and disinfection. Drinking-water supplies should also be protected from contamination during distribution. |
| 5. | *Cryptosporidium e.g. C. hominis C. parvum.* | *Cryptosporidium* oocysts have been detected in many drinking-water supplies. Concentrations of oocysts as high as 14 000 per liter for raw sewage and 5800 per liter for surface water have been reported. | *Cryptosporidium* generally causes self-limiting diarrhea, sometimes including nausea, vomiting and fever. | The oocysts are extremely resistant to oxidizing disinfectants such as chlorine, but UV light irradiation inactivates oocysts. Membrane filtration processes can also be used. |
| 6. | *Giardia* spp *E.g.: G. intestinalis*, also known as *G. lamblia* or *G. duodenalis.* | Contaminated drinking water, recreational water and, to a lesser extent, food have been associated with outbreaks. | Diarrhea and intestinal malabsorption. | *Giardia* cysts are resistant to chlorine, prevention of source water contamination by human and animal waste, followed by adequate treatment and disinfection and protection of water during distribution. |

of the infected humans or animals. A similar route may be observed with viruses such as enteric virus, which affects the gastrointestinal systems of humans, leading to acute and chronic infections. Some viruses that cause respiratory infections in humans can spread through respiratory droplets, which may also be excreted in the feces of the infected person. This shows that the pattern and route of infection can be varied for different microbial pollutants and their sub types.

One of the most common causes of disease and infections due to contaminated water is the presence of protozoa. This is considered to have a major public health and socioeconomic impact and is a major concern worldwide. For instance, cryptosporidiosis is an infection caused by *Cryptosporidium* species that are present in drinking water. This is a major issue faced by many parts of the world. The main challenge in treating water containing protozoa is the presence of oocysts, cysts and eggs, which are often resistant to typical disinfectants that are used, and, in some instances, they may be difficult to remove by filtration processes too.

Yet another type of pollutant is the helminths or worms such as roundworms, flatworms and hookworms, which are a common source of infection in humans and animals globally. Some of these are transmitted through contaminated water or are associated with using untreated waste water for agriculture. For example, *Dracunculus medinensis*, commonly known as "guinea worm", is the only nematode associated with significant transmission by drinking water. Guinea worm infections are generally found in countries of sub-Saharan Africa. The only source of this infection appears to be drinking water, containing infected cyclops, and it is often prevalent in rural areas where piped water may not be available.[3]

### 1.3.4 Emerging Contaminants

Over the last few decades, with increasing industrialization and human and anthropological activities, several classes of pollutants have been found in drinking water. Most of these contaminants are discussed in the earlier section. However, there are innumerable pollutants, still present in drinking water, that usually do not fall in the categories of physical, chemical or microbial pollutants. These substances are called "emerging contaminants" and pose an equal challenge as the other pollutants in drinking water treatment. Typically, they are denoted as Contaminants of emerging concern (CECs) by the Environmental Protection Agency (EPA) and they include pharmaceuticals and personal care products (PPCPs).

The PPCPs are often found in surface waters in low concentration and they may have a detrimental effect on the marine life. Most of the drugs that are taken orally pass through the human body and are excreted and finally may find their way into the water supply, especially if there are possibilities of sewage contamination. Similarly, personal care products such as soaps, fragrances and cosmetics also land up in the water supply. Many of these substances are known to act as endocrine disruptors (EDCs). EDCs are compounds that change the normal functions of hormones, resulting in myriad heath issues such as altered reproductive effects in aquatic organisms. They may be toxic, leading to severe acute toxicity issues. Since these EDCs are different in nature, conventional testing procedures that are adopted for other pollutants may not be sufficient to detect these compounds.

Owing to this challenge, the EPA developed a document titled '*White* Paper Aquatic Life Criteria for Contaminants of Emerging Concern: Part I Challenges and Recommendations' that outlines the technical issues and recommendations that can be used to modify the existing guidelines. This was primarily published to address the emerging contaminants and come up with a better-quality criterion for drinking water with the current resources and knowledge available. Emerging contaminants are important due to the risk they may pose to human health and the environment, which has not been fully elucidated yet. Globally, trace amounts of these emerging contaminants are being detected in the water supply in many regions and the EPA is working towards enhancing its understanding of these pollutants.[4]

Several experts worldwide are investigating the concentration, types, fate, behaviors and toxicity levels of these emerging contaminants and reports from such studies clearly point out that there is a dearth of published health risk standards for the emerging contaminants that can provide a suitable guideline in treating them. Moreover, several new emerging contaminants are being introduced into the environment and therefore into drinking water that have not yet been detected. Therefore, superior detection methods and treatment technologies are needed in the years to come to detect and remove these contaminants from potable water systems, and at the same time, appropriate measures to prevent the introduction of new contaminants are the need of the hour.

## 1.4 Water Quality Guidelines

Drinking water should be safe and clean when it reaches the end user, and to ensure this, it is necessary to have water quality guidelines in place so that the water quality can be compared with these guidelines before it is consumed. The World Health Organization (WHO) has played the pivotal role of setting the water quality guidelines, which have been considered the standard for ascertaining drinking water quality globally. As per these guidelines, safe drinking water may be defined as water that does not cause any significant health risk to the end user over the lifetime of consumption.

The WHO guidelines are mainly used to support the development and implementation of different risk management methods that ensure safe drinking water supplies do not contain any hazardous constituents that may pose a health issue. These guidelines were mainly envisioned to provide a scientific basis for the development of various national and regional standards worldwide. Moreover, the WHO guidelines describe the minimum requirement of safe practice to provide safe and clean water to the consumers and most importantly, these guidelines state the numerical guideline values for different acceptable and unacceptable constituents present in water.

It is important to understand that there are no unique drinking water standards that are universally applicable in all countries globally. This is mainly because the needs and capacities of water treatment of one country may vary significantly and the approach of one country or region towards drinking water standards may not necessarily translate to another country or region. This is one of the main reasons for not encouraging the implementation of international standards for drinking water quality. The purpose of the WHO guidelines is to enable countries and regions to form national and regional drinking water standards that can be readily implemented in the respective regions and protect public health.

For instance, the WHO guidelines broadly describe and state that the greatest microbial risk is due to the consumption of water that is contaminated with feces from human or animals/birds, since a plethora of pathogenic bacteria, viruses, protozoa and helminths can find their way into the human body through contaminated water. Therefore, control of microbial contamination should be the prime goal and must never be neglected by water authorities. Chemical contaminants, on the other hand, usually lead to a health threat only after prolonged periods of exposure. Yet another important aspect of water treatment stated in the WHO guidelines is that disinfection should never be compromised when endeavoring to control disinfection byproducts.

Important pointers such as these can be easily adopted by different countries and regions worldwide while preparing the national and regional drinking water standards. A preventative integrated management approach that ensures cooperation from all agencies and stakeholders is the best method to ensure drinking water safety. This will ensure that the quality and safety of the potable water produced are comparable with the WHO standards and at the same time compatible with the local guidelines as the water source, its constituents, treatments methods and needs of the consumers vary from one place to another.

For example, in the US, the safe drinking water act (SDWA) is the main central law that safeguards the drinking water quality and under this law, the Environmental Protection Agency (EPA) sets the drinking water quality standards. This is done by setting the limits for several contaminants routinely found in drinking water. These limits are usually based on the levels that protect human health and should be achievable by employing the best available water treatment technology. Additionally, the water quality testing methods and the time intervals to schedule these tests are also decided by the EPA. It also supervises the states, regions and various water providers who implement those standards. Some examples of pollutants and their limits as per the EPA are listed in Table 1.3.

**Table 1.3** Some water pollutants and their limits as per the EPA.

| S. no. | Pollutant | Maximum Contaminant Level Goal (MCLG)[a] $mg\,L^{-1}$ |
|---|---|---|
| 1. | Uranium | 0 |
| 2. | Xylenes (total) | 10 |
| 3. | 1,1,1-Trichloroethane | 0.2 |
| 4. | Styrene | 0.1 |
| 5. | Lindane | 0.0002 |
| 6. | Chlorobenzene | 0.1 |
| 7. | Acrylamide | 0 |
| 8. | Nitrite (measured as Nitrogen) | 1 |
| 9. | Lead | 0 |
| 10. | Fluoride | 4 |
| 11. | Arsenic | 0 |
| 12. | Bromate | 0 |
| 13. | Viruses (enteric) | 0 |
| 14. | Total Coliforms (including fecal coliform and *E. coli*) | 0 |
| 15. | *Cryptosporidium* | 0 |
| 16. | *Legionella* | 0 |

[a]As per EPA, Maximum Contaminant Level Goal (MCLG) means the level of a contaminant in drinking water below which there is no known or expected risk to health.

## 1.5 Need for Water Treatment

It is evident from the previous discussions on the spectrum of pollutants that contaminated source water can lead to various health issues. This is especially a major challenge in rural regions of developing countries as there are higher chances of contamination due to poor sanitation and hygiene practices and the low socio-economic conditions. It is well established that fecal contamination of water supplies prevails in many rural regions of developing countries. Over 2 million infections per year that cause almost 600 000 deaths primarily in developing countries are reported by the WHO.

Among the potential diseases due to contaminated drinking water that particularly affect developing countries are bacterial infections due to Shigella species that can be triggered even by a low count of 10 to 100 organisms. Similarly, enteropathogenic *E. coli* infections are also known to occur in developing countries with infants being affected the most. Viruses such as Hepatitis A virus and protozoa such as *Blastocystis hominis* are probably the most common pathogens found in water sources in developing countries owing to the poor hygienic conditions. Similarly, Schistosomiasis, which is caused by parasitic worms (*e.g. Schistosoma mansoni*), is also a great public health concern globally especially in developing countries where it is endemic.

Due to these factors that cause human health hazards that can even be fatal, safe and clean drinking water is of paramount importance. One way to ensure this is preventing contamination of the source water so that the raw water is clean and its quality is within the desirable WHO or the respective

National standards. However, this is not always possible as most of the time, especially in developing countries, contamination of the source water occurs by mixing of sewage water and due to general poor hygiene and sanitation around the drinking water source/storage. In some cases, waste water from industrial effluents and domestic waste can lead to the introduction of undesirable and deleterious pollutants. Therefore, water treatment is the only option to ensure safe drinking water.

Drinking water treatment is the main solution that can inactivate or eliminate the various types of pollutants that may be present in the raw water. Any water treatment method will involve one or more steps to remove the pollutants and render the water clean and fit for consumption. Generally, pre-treatment such as flocculation, sedimentation and pre-filtration is carried out, which removes most of the coarse particles and turbidity that may be routinely found in water. Subsequently, different treatment methodologies such as filtration by sand, membranes, ceramics or activated charcoal, and exposure to UV or solar irradiation may be used depending on the source water quality and the required end use. If microbial pollutants are present in the source water, disinfection is of prime importance.

Disinfection, as the name suggests, is the inactivation and/or removal of pathogenic microorganisms from contaminated water. This is especially important when the surface or groundwater is subjected to fecal contamination. Disinfection may be physical, chemical or hybrid in nature and most commonly, the latter is used. Chemical agents such as chlorine and ozone are routinely used in water treatment to kill the infectious microbes that may be present. These chemical disinfectants may not necessarily inactivate all the microbes, *e.g.* chlorine may not be as effective on Protozoans like *Cryptosporidium* and some viruses. It is also important to understand that the disinfectants may not be very effective if the source water is highly turbid and when the microbes are surrounded by particulate matter leading to clumping and the formation of agglomerates.

Moreover, chemical disinfectants are known to form disinfection byproducts due to their reaction with organic matter present in the source water. Many disinfection byproducts like Trihalomethanes are potentially carcinogenic and may lead to health risks, if present in excess amounts in the final disinfected water. However, it is necessary to reemphasize here that at no point should the disinfection of water be compromised for the sake of the potential formation of disinfection byproducts. Chemical disinfection of only the source water may not be enough to make the water completely safe as contamination maybe reintroduced in the piping systems and distribution networks. Therefore, some amount of residual chemical disinfectants is added to ensure the safety of drinking water when it reaches the consumer. Needless to say, storage vessels used by the end user and source water protection are also required to ensure that the drinking water is free of contamination.

Water treatment in developing countries is generally carried out by using low cost methods that are easy to operate, involves the use of locally available materials and is feasible either at the point of use at household level or at a community or larger scale level as the case may be. Disinfection in developing countries primarily depends on the use of chemical disinfectants such as commercially available bleach or a sodium hypochloride solution. This is attributed to the fact that it is easily available, relatively safe to handle, inexpensive and easy to apply even at the household level. Therefore, techniques such as UV disinfection are not generally useful for rural set-ups in developing countries due to the need for reliable electricity, cost of the UV set-up and its maintenance requirement.

Thus, based on the previous discussion, the need for clean and safe water, on one hand, and the possibility of various physical, chemical and biological pollutants that may be present in the source water cannot be overstated. Therefore, drinking water treatment techniques play a pivotal role globally, especially in developing countries. This is the main focus of this book and an attempt has been made to cover most of these aspects to highlight the current scenario globally. The following section gives a brief overview of what this book has to offer.

## 1.6 Objective of This Book

As the title of this book and the preceding sections of this chapter suggest, the focus of this book is elaborating various drinking water treatments for removing or eliminating different pollutants present in source water that are well suited for developing countries. Based on this, each chapter deals with a topic leading to conclusions on this topic. Chapter 2 deals with drinking water treatment techniques that are feasible in rural set-ups. It includes descriptions of and discussion on some of the simple yet effective and economical technologies that have been conventionally used. Novel and upcoming methods that are evolving around the globe to cater to the ever-increasing potable water needs are also highlighted. It will indeed be interesting to find out more about individual and collective efforts from various stakeholders including research scientists and experts in the field of water that have immensely contributed to ensuring safe and clean drinking water is accessible to the remotest areas worldwide.

In a very different yet fascinating way, Chapter 3 will throw light on a simple and beautiful way of conserving water by collecting rain water, which is a gift of nature. Techniques of using the harvested water after appropriate treatment for potable and domestic purposes will clearly show how humans can make efforts to save and use rain water effectively. Along similar lines, Chapter 4 will reveal fascinating ways of reusing waste water after treatment for potable and non-potable purposes. This methodology of conservation of water for reuse is where the world is headed and will form one of the major ways of ensuring safe and clean water for humankind in the decades to come.

As the heart of this book is on drinking water treatment, the explanation of any technique would be incomplete without appropriate examples. Thus,

Chapter 5 illustrates a few of the treatment methods described in this book by way of specific case studies. These case studies are included essentially to demonstrate the economic application of the said water treatment techniques in rural scenarios of developing countries. The success of these methodologies as depicted in the case studies gives a ray of hope for the brilliant possibilities for providing safe and clean drinking water to the rural regions of even low-income countries.

For any book that involves discussion on a plethora of water treatment techniques for myriad pollutants that may be present in water, a comparison of the same is of utmost importance. Therefore, Chapter 6 reveals to the reader the best possible treatment method for any contaminant in a developing country scenario. In other words, the methods described in this book will be compared with respect to efficiency in pollutant elimination or inactivation, cost, scalability and overall feasibility and ease of deploying in developing countries. Chapter 7 will essentially wrap up the entire book in a nutshell and include a summary, conclusions and future recommendations drawn based on all the previous chapters.

## References

1. http://www.undp.org/content/undp/en/home/sustainable-development-goals/goal-6-clean-water-and-sanitation/targets/ (last accessed July 2018).
2. http://www.who.int/mediacentre/news/releases/2017/launch-version-report-jmp-water-sanitation-hygiene.pdf?ua=1(last accessed July 2018).
3. Guidelines for drinking-water quality: fourth edition incorporating the first addendum. Geneva: World Health Organization; 2017. Licence: CC BY-NC-SA 3.0 IGO.
4. https://www.epa.gov/sites/production/files/201508/documents/white_paper_aquatic_life_criteria_for_contaminants_of_emerging_concern_part_i_general_challenges_and_recommendations_1.pdf (last accessed July 2018).

CHAPTER 2

# *Treatment Techniques for Developing Countries*

The previous chapter introduced the need for clean and safe drinking water and highlighted that its availability is the most important requirement for any individual, community or nation. This was drawn from crucial information based on the United Nations report that pointed out the grim state of water scarcity that would affect the world by 2025. According to the latest update on the SDG goal for drinking water released by JMP, it is distressing to note that there are many people who still lack access to safe drinking water especially in rural areas. Given the myriad chemical, physical and biological pollutants that were discussed at length in the previous chapter and their detrimental health effects on humans, it is imperative to have appropriate disinfection techniques that can effectively remove the contaminants or render them inactive (in the case of microorganisms). Therefore, drinking water treatment for developing countries is one of the vital aspects to be considered and this chapter will address this central topic.

## 2.1 Introduction

Drinking water treatment is very important for getting safe potable water to meet human needs in any region globally and this is especially critical in the developing countries. This is mainly because low income nations cannot afford expensive technologies for drinking water treatment and the lack of electricity in most of these areas is yet another deterrent in the application of routine water treatment techniques. Therefore, only those methods that are simple, easily adoptable, inexpensive and user friendly can be envisaged as useful methodologies to get safe and fresh drinking water in the developing countries. Some of these methods are solar disinfection (SODIS), filtration, herbal treatment and desalination. Additionally, a plethora of novel

Drinking Water Treatment for Developing Countries: Physical, Chemical and Biological Pollutants
By Aniruddha Bhalchandra Pandit and Jyoti Kishen Kumar

Published by the Royal Society of Chemistry, www.rsc.org

emerging techniques that are simple and cost effective are also available for water treatment in the developing nations.

Solar water disinfection can be used in any country that has sunny weather most of the time during the year such as in regions located near the equator. This technique is very ancient but over the last decade, a lot of research has led to the development of innovations in SODIS such as the use of combinations with photo catalysis, thermal treatment, use of natural additives and modifications in the Polyethylene terephthalate (PET) bottle. Efforts to scale up SODIS for large scale applications has resulted in its application at community levels.

Filtration, which is almost always the choice of water treatment in most cases, is also easily adaptable in developing countries as it is simple and easily operated at household levels. Ancient methods of slow sand filtration are still applicable and found to be very effective in developing countries. Apart from these methods, rapid sand filters, ceramic or fired clay earthen filters and the use of natural filters have evolved over the last few decades and have been found to be very useful to treat water to near potable standards in the developing nations. Since most of these countries usually have a rich collection of medicinal plants, herbal approaches to treat water have also resulted in encouraging results. Plants such as *Moringa oleifera* (drumstick), which are indigenous to many developing nations, have been found to effectively treat water and render it potable.

Yet another fascinating method is desalination of sea water to cater to the needs of a developing nation. Several desalination plants operate globally, which could be an answer to the ever-increasing demands for drinking water. This is especially useful to developing countries when the energy for the operation of the desalination process is obtained from renewable energy sources such as solar, wind, wave and geothermal energy as these energy sources are easily harnessed in developing nations and result in inexpensive desalination methods that can be easily adopted to cater to the needs of the rural population.

Over the years, despite many advances in the techniques mentioned here, there has always been a dearth of newer simple methods specific for low-income regions. To answer this issue, several experts, groups of scientists and thoughtful individuals have come up with novel techniques such as the solar ball, LifeStraw, drinkable book, water purifying bicycle and many more that are discussed at length in this chapter.

## 2.2 Treatment Methods

### 2.2.1 Solar Disinfection (SODIS)

Sun, a natural resource, has been harnessed by humans for eons. Our ancestors used solar rays to sterilize and disinfect water although they may not have clearly understood its mechanisms. Since solar energy is easily available at no cost, it has enthused the interest of scientists to explore the possibility of employing this powerful source for disinfection of water especially in developing countries. In low income countries where the economic conditions are poor, SODIS is an excellent means of obtaining safe drinking water if the region has a sunny climate.

It is indeed remarkable to note that more than 2 million people in 28 developing countries use SODIS for daily drinking water treatment. It has been suggested that exposing water in bottles or plastic bags to sunlight can effectively disinfect the next day's supply of water after a natural disaster and this is an easy method that can be adopted by the local population without any special assistance. It is a well-established method and a lot of NGOs, state and national stake holders are already promoting SODIS, and this technique has been in practice for a long time. There is ample literature available on this topic owing to the research and development work initiated by various scientists to enhance this methodology and make it more efficient. Therefore, this section will only highlight the main aspects to SODIS.

The method of using solar disinfection is extremely simple and involves pouring contaminated water into plastic bottles and exposing them to sunlight for 6 hours. This results in the inactivation of most microorganisms that may be present in water. The mechanism of inactivation has been studied extensively and can be summarized as one or a combination of three possibilities, *viz.* (i) the microorganisms present in water absorb the UV rays of the solar radiation and get inactivated due to DNA damage; (ii) the UV rays can also bring about an indirect effect by exciting photosensitizers (pigments, porphyrins) within the microbial cell, which in turn activates type I or type II mechanisms.

The former occurs at a cellular level and the latter is an action of molecular oxygen. Both the mechanisms lead to the formation of reactive oxygen species (ROS) such as hydroxyl radical, hydrogen peroxide and superoxide. The ROS damages the cell by acting on the cellular DNA, cell membrane and proteins. (iii) The third mode is by the thermal effect of solar radiation that essentially involves infrared radiation. In this case, the absorption of IR radiation results in the increase of temperature of the water being treated that inactivates the microorganisms by a process called solar pasteurization.[1] Due to increasing demands for safe water, there has been a need to modify and improve the traditional SODIS technique. This led to the development of numerous variations of SODIS that are discussed in the following sections emphasizing their application in different developing countries.

#### 2.2.1.1 Conventional SODIS

The SODIS technique is very simple and easy to use, as shown in Figure 2.1. The first step is to clean a transparent PET bottle with a mild detergent to ensure that the glass surface is free of dirt and grime. This is important especially if the bottle is being used for the first time. The water to be treated is filled into the PET bottle preferably after filtering through a muslin cloth

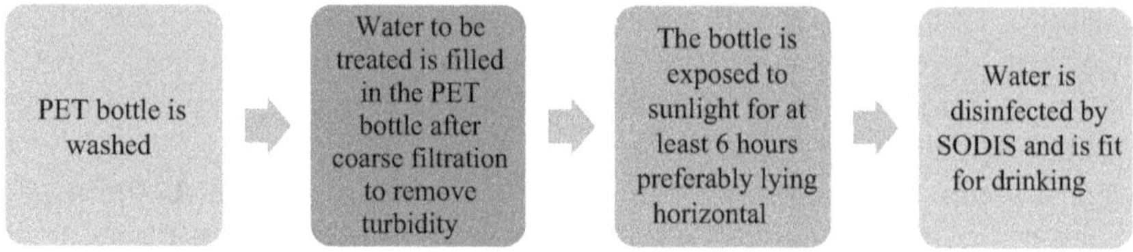


**Figure 2.1** The SODIS method.

to remove turbidity and placed in sunlight for at least 6 hours. After the exposure to solar rays, the water is disinfected and safe for human consumption. Although the methodology is simple, there are certain factors that need careful consideration.

PET and clear plastic bottles are recommended as they are light and unbreakable. Moreover, they are easily obtained in almost all regions. Needless to state, the bottles must be colourless and transparent to ensure the penetration of solar radiation. It is important to note that solar radiation intensity is reduced by increasing turbidity and water depth. Therefore, raw water to be treated must have low turbidity (<30 Nephelometric Turbidity Unit [NTU]). Similarly, the water depth in the bottle should be small and not exceed 10 cm to facilitate sufficient solar rays to reach the water, thus it is recommended to place them horizontally.

Yet another important aspect to be considered is the cloudiness of the sky. It is recommended that if more than half of the sky is covered with clouds, the bottle must be placed in the sun for 2 consecutive days. If the sky is clear or less than half of the sky is clouded, 6 hours is sufficient to disinfect the water. Organic compounds such as humic acids and inorganic substances such as salts present in water also affect the overall effectiveness of SODIS. While organic compounds may either increase or decrease the SODIS effect, inorganic compounds usually act synergistically with SODIS. Similarly, dissolved oxygen also enhances SODIS by increasing the reactive oxygen species production.

Yet another factor that affects SODIS is the incident light quality and intensity. Generally, the UVA and UVB region of the solar radiation spectrum (290–350 nm) leads to the inactivation of microbes. Intensity of above 500 W m$^{-2}$ leads to an effective SODIS and temperatures of above 40–50 °C are known to enhance the SODIS effect. Most importantly, the type of microorganism to be inactivated and the growth phase in which it exists are important factors while applying SODIS as this would determine the duration and solar radiation dosage required to bring about water disinfection. For instance, generally, 10 mins of SODIS at around 56 °C is enough to inactivate cysts whereas fungi need 3–6 hours of SODIS depending upon the species.

Bacteria such as *E. coli* usually exhibit an initial delay of 0.5 to 2 hours after which the solar inactivation is exponential. Viruses are more resistant than bacteria and usually require twice the dosage of solar rays for complete inactivation. A range of fungi, including *Aspergillus niger, Aspergillus flavus,*

*Candida* sp. and *Geotrichum* sp., and pathogenic microorganisms like *Salmonella typhi* and *Shigella flexneri*, several yeasts and moulds can also be inactivated by solar disinfection. Most bacteria and fungi can be inactivated within 8 hours of SODIS at 870 $W m^{-2}$ and *C. parvum* oocysts can be effectively eliminated with a solar radiation dosage of ≥600 $W m^{-2}$ for a duration of 4 to 12 hours. Protozoa such as *Acanthamoeba*, *Entamoeba*, *Naegleria* and *Giardia* can also be inactivated with a solar intensity of 550 $W m^{-2}$ for 6 hours.

Studies on SODIS have been reviewed in the literature and a comprehensive list of various instances where SODIS has been used to inactive microbial contaminants is amply described.[2,4] Therefore, this section discusses recent examples of SODIS used for treating contaminated water. An interesting initiative worth mentioning here would be Eawag, the Swiss Federal Institute of Aquatic Sciences and Technology that supports SODIS and helps its promotion worldwide along with its NGO partner, Helvetas Swiss Intercooperation. Together, they have carried out several projects in different developing countries such as the promotion of SODIS and water treatment at the household level in Benin, Africa. This initiative gave around 66 000 people access to safe drinking water and has resulted in the local government adopting SODIS as a means for safe potable water.[3]

In another project based on using sunlight to treat drinking water in Vietnam, Asia, SODIS was shown as an effective method to get pure and safe drinking water, and it caused the national government to integrate the SODIS method into a national program funded by the state. It also led the local authorities to adopt guidelines related to the handling and storage of clean drinking water at the household level. Yet another project was initiated in the highlands of Bolivia, Latin America to teach the people the importance of clean water in the school curriculum. It was thought that teaching the basic relation between hygiene, drinking water and health to local women's groups and teachers would have a huge impact on the entire community.

Several studies in the last few years have reinforced the fact that SODIS can effectively prevent myriad water borne diseases like cholera and diarrhea. SODIS can be effectively used during a cholera outbreak to prevent the disease spreading and death due to diarrheal dehydration. Thus, SODIS is an adequate intervention and can be used in lower and middle-income countries to reduce diarrhea, disease transmission and associated deaths. Yet another interesting study on childhood diarrhea rates in sub Saharan Africa showed that improved bags made of polyethylene (PE) with ethylene vinyl acetate (EVA) were good substitutes for the conventional PET bottles for carrying out SODIS.

Numerous investigations have been carried out on the type of microorganisms that are inactivated and viruses are one among them especially owing to the increasing incidences of water borne viral infections worldwide. In one such study, inactivation of bacteriophages and human viruses by SODIS was investigated by taking sample waters from different regions of

Switzerland and India. The findings of this study showed that SODIS could reduce the load of echoviruses at high temperatures whereas further complimentary measures were needed to inactivate certain viruses resistant to oxidation such as the bacteriophage. This clearly shows that the basic SODIS technique requires enhancement by various means to effectively remove all water pathogens, which is described in the following section.

### 2.2.1.2 Thermal Enhancement of SODIS

Due to the synergy between optical and thermal inactivation of microorganisms that has been shown in various studies especially above >45 °C, various attempts have been made to enhance SODIS with thermal effects. Some of these techniques include the use of absorptive materials and painting the PET bottles black from the outside so that the black surface increases the absorption of solar radiation thereby accelerating SODIS efficiency. Solar concentrating collectors as well as reflectors can also aid in increasing the thermal effect but not as much as a blackened surface. However, painted bottles are not useful on cloudy days as the absorption of solar radiation is insufficient for practical use.

It is generally known that most economic solar collectors have an average conversion efficiency of around 30% and they can collect solar rays even on cloudy days. It should be noted that in winter, the system would not reach the desired temperature because of heat losses to the environment. Hence, there have been attempts to pair a solar thermal collector to a radiation collector device to tackle this issue. Although efficient solar thermal collectors are available, their high cost appears to be a major limitation in their application as a part of a SODIS system for use in developing countries.

### 2.2.1.3 SODIS and Photocatalysis

Titanium dioxide ($TiO_2$) is a photocatalyst that can accelerate SODIS and has been widely investigated for water treatment applications. Most of the research work reports the use of $TiO_2$ either as an aqueous suspension as fine nano or microparticles or immobilized on a suitable solid substrate. The latter case essentially consists of a flat plastic sheet that is coated with $TiO_2$ on the upper surface and this is subsequently inserted in the PET bottle and subjected to SODIS. This technique is particularly useful in developing countries, for instance, in a rural community of the State of Chihuahua, it was observed that the water quality improved by complete elimination of coliform bacteria making the water free of pathogens and fit for human consumption.

In yet another study, SODIS reactors fitted with flexible plastic inserts coated with $TiO_2$ powder were found to be much more effective against *Cryptosporidium parvum* oocyst than those that were not coated, indicating its promising application in household water disinfection. However, both (suspension and immobilized) methods have pros and cons. While the suspension form results in increased surface area for reaction, it also poses problems in recycling of the $TiO_2$. On the other hand, the immobilized version facilitates its reuse but also decreases the effective catalyst surface area available for reaction to occur. Moreover, costs of immobilization support materials also add to the cost of the technique. Therefore, careful consideration is needed to select either free or immobilized titanium dioxide to use as a catalyst to enhance SODIS.

There are several mechanisms of action suggested by different researchers for the inactivation of microorganisms by $TiO_2$. Some propose that the ROS species of hydroxyl radicals are primarily responsible for inactivation. Other studies also include hydrogen peroxide and oxygen radicals as contributors in the inactivation process by way of attacking the DNA, RNA or cell membrane of various susceptible microbial species. Some even suggest the direct contact of the catalyst with the bacterial cell membrane causing it to rupture.

Hosts of microorganisms such as *Clostridium perfringens* spores, *E. coli*, *Cryptosporidium parvum* and a plethora of bacteria and viruses have been successfully inactivated by using $TiO_2$ in combination with SODIS. However, the major limitation of the process appears to be the post treatment recovery of suspended titanium dioxide as the catalyst, as if it is present, it can lead to taste issues in water rendering it unpalatable. Also, there are insufficient studies that have assessed the health hazards associated with the presence of high levels of the catalyst in treated water.

Overall, it can be concluded that SODIS is accelerated by way of a reduction of the irradiation dose or reduction in the duration of sunlight exposure by using titanium dioxide in combination with SODIS. This combination also extends the range and type of microbes that SODIS can inactivate. Most importantly, addition of $TiO_2$ can make the SODIS process continuous. Yet another interesting modification of photocatalysis enhanced SODIS is electric field enhancement on the catalyst surface. This boosted the production of reactive oxygen species such as hydroxyl radicals and thus accelerated SODIS. This technique was called photoelectrocatalysis and was researched extensively to elucidate its disinfection capability. The studies revealed that it could successfully inactivate many resistant microbes such as *Clostridium perfringens* and their spores and cysts.

### 2.2.1.4 SODIS and Additives

Yet another technique of improving SODIS is by adding various additives that are chemical or natural in origin. The main goal is to enhance the efficiency in inactivating microbes. Among the vast investigations that have gone into exploring such additives, simple dyes were initially experimented with SODIS. Dyes such as methylene blue and rose Bengal when added to water and subjected to SODIS resulted in enhanced production of various reactive oxygen species and resulted in higher inactivation rates. The major concern was the consumption of water containing these dyes. This was eventually addressed by adding chlorine to decolorize the dye prior to the consumption of the treated water, although solar radiation can also fade colours and therefore reduce the colour given by the dye. However, this process was found to be more appropriate for waste treatment due to its impracticality in the field conditions for potable water production in developing countries where the availability of the two chemicals, *viz.* the dye and chlorine, was relatively difficult.

Other chemicals that have been studied include hydrogen peroxide (100–1000 μM), ascorbate (25 to 37.5 μM), copper (2.5 to 7.7 μM) and sodium percarbonate (125 $mg L^{-1}$) that successfully accelerated the SODIS inactivation of a spectrum of microorganisms such as *coliforms*, *MS2 coliphages* and several *Enterococcus* species. Riboflavin has also attracted some interest in its use as an additive in conventional SODIS experiments. This is mainly because Riboflavin enhanced SODIS can not only kill coliforms but also protozoal cysts such as *A. polyphaga* cysts and *Acanthamoeba castelleni* cysts. Not all the chemical additives are economical and readily available given the fact that SODIS is particularly targeted at low to medium income populations. This triggered the search for natural additives that are inexpensive and easily obtainable in the local area where SODIS is practiced.

A host of natural ingredients ranging from food preservatives such as lime and vinegar to household spices and common plant additives have been explored in the last few years. Among the natural additives, those based on citrus fruits have been researched extensively and many scientists have proved that SODIS was accelerated by the presence of these additives as they found enhanced inactivation rates of bacteria such as *E. coli* and *Enterococcus* sp. as well as viruses such as MS2 coliphage. SODIS combined with lime juice or lime pulp could reduce *E. coli* levels within half an hour of treatment. Literature reports described *E. coli* reductions of ~6 log units for SODIS and lime bottles compared to 1.5 log units for standard SODIS bottles in the same time (30 min). It is necessary to note that such a treatment duration is on a par with boiling and other household water treatment methods.

In yet another interesting work on the enhancement of the efficacy of SODIS with additives, aqueous extracts of plants such as *Cassia alata*, *Cassia occidentalis*, *Carica papays*, *Phylanthus niruru*, and *Coleus kilimandschari* were found to inhibit microorganisms in water by a photosensitization process within two hours of exposure to sunlight. Similarly, essential oils of *Citrus bergamia*, *Citrus limonum* and *Citrus reticula* used as photosensitizers with SODIS completely inactivated fecal coliforms present in water within 2 hours of sunlight exposure.

A photosensitizer is essentially an active molecule in a substance that has one or more chromophore groups. These chromophore groups are basically functional groups in a molecule that have the capability of absorbing visible light. Thus, photosensitization is a method using an active substance, called the photosensitizer, which enhances the disinfectant action of sunlight. This process is very similar to photocatalytic oxidation, but it offers an advantage over the photocatalytic process because the sensitizer absorbs light in the visible spectrum unlike most of the photocatalysts, which require light at a

UV wavelength. This results in allowing the use of a greater percentage of available sunlight. Thus, using the sunlight as the energy source and the dissolved oxygen in the water as the oxidizing agent, the photosensitization process was reported to inactivate bacteria by different mechanisms.

The mechanism of action is either a type I mechanism where the photosensitizer interacts with a biomolecule to produce free radicals or a type II mechanism where singlet oxygen is produced as the main species responsible for cell inactivation. Thus, initially, there is a preferential accumulation of the photosensitizer in the microbial cell by electrostatic interactions or diffusion, followed by targeted illumination that induces photochemical activation of $O_2$ into its excited state. Subsequently, this highly reactive oxygen species reacts with macromolecular targets within the bacteria leading to their inactivation.

It is important to note that there are various types of photosensitizers such as chemical and natural varieties. All of them have the capacity to absorb light in the UV–visible spectrum and are capable of producing singlet oxygen. Some of the chemical types are the dyes such as methylene blue, Bengal rose, eosin, porphyrins, *etc.* Among the natural photosensitizers, plants that are used traditionally to treat microbial and parasitic infections are also capable of producing singlet oxygen. This is attributed to the fact that these plants contain certain molecules that are responsible for their colouration and secondary metabolites. These molecules, also called chromophores, when present in water and in the presence of energy in the form of sunlight, could absorb the energy and pass from ground state to the excited state. Subsequently, this stored energy is passed on to the oxygen present in water, which in turn moves from the triplet state to the singlet state. This is responsible for the inactivation of microorganisms.

It is interesting to note that the initial studies of water disinfection by photosensitization were carried out almost four decades ago using methylene blue as the sensitizer and sunlight and UV were utilized as the energy source. Fascinatingly, a reduction of the order of $1.3 \times 10^9$ CFU *fecal coliforms*/100 ml was observed for a concentration of about 0.5 mg $L^{-1}$ methylene blue dye.[10] Similar observations were made by other experts, for instance in one water disinfection study, it was observed that methylene blue yielded a substantial reduction in the population of *E. coli* (reduction of $10^3$–$10^4$ CFU per ml for *E. coli* at a concentration of 10 mg $l^{-1}$).[11]

Among the several plants used in traditional medicine, leaves of *Cassia alata*, *Cassia occidentalis*, *Carica papaya*, *Coleus kilimandschari,* and *Phyllanthus niruri* are used in the treatment of microbial and parasitic infections.[12] In one study, the researchers found that water treated with plant extract and exposed to sunlight inactivated all the pathogenic microorganisms within 2 hours.[13] The results showed that an initial *fecal coliform* count of $4 \times 10^4$ CFU per ml reduced to 0 CFU per ml (4 log reduction) when water was treated with *Citrus bergamia* (1 ml $liter^{-1}$ of water) and exposed to sunlight for one hour. *Citrus limonum* (1 ml $liter^{-1}$ of water) and *Citrus reticula* (1 ml $liter^{-1}$ of water) were found to be comparatively less effective and brought down the fecal coliform count to $2\times10^3$ and $3\times10^3$, respectively, (one log reduction) at the end of one hour of treatment.

Since better results were obtained when water was treated with *Citrus bergamia* and exposed to sunlight, the researchers carried out further studies to investigate the disinfection activity of *Citrus bergamia*. This was essentially done by testing its inactivation efficacy of two microorganisms, *fecal coliforms* (Gram−) and *fecal enterococci* (Gram+). It was observed that fecal enterococci were more sensitive to the disinfection of water by photosensitization than fecal coliforms. This was mainly attributed to the differences in their cell membranes. In the case of fecal enterococci, the polylipopolysaccharide layer was absent in the cell membranes, leading to relatively easier damage to the cell membranes by the singlet oxygen. Fascinatingly, no regrowth was noticed after the photosensitization phase for both *fecal coliforms* and *fecal enterococci* even after 24 hours.

The plant extracts used in the study were found to contain sensitizers that were similar to methylene blue and were capable of inactivating microorganisms in water. Moreover, it was observed that the activity of the photosensitizer molecules in the plants was enhanced considerably in the presence of oxygen. Further screening of the extracts revealed that the molecule responsible for the photosensitization activity in the *Citrus* was found to be Methoxy-5 psoralen (MOP-5), a common coumarin commonly used in the treatment of psoriasis, a skin disease. The mechanism of action of MOP-5 is like the activity of the other photosensitizers discussed earlier. In the presence of light, MOP-5 undergoes photo-excitation and transmits its energy to oxygen dissolved in water, which moves from its fundamental triplet state to the singlet excited state. The singlet oxygen thus generated leads to inactivation of coliforms and enterococci in water.

Even though MOP-5 was found to be the active component responsible for the inactivation of microbes, it was observed that the photosensitizer remained in solution even after disinfection, thereby making the treated water unfit for consumption. Therefore, there was a need to remove the photosensitizer from the solution after treatment, and one way to do this was by immobilization of the photosensitizer on a solid support. Immobilization of the photosensitizer on a solid support would not only assure its removal from the solution after the treatment, but it is also an economical and environmentally friendly option.

Since the active ingredient is bound to give better disinfection as compared to the whole plant extract, MOP-5 was further explored for its efficacy and reusability by attaching it on two solid supports made of polystyrene (designated as A and B in the study). Friedel–Crafts acylation was used for the immobilisation of MOP-5 on the support. It is interesting to note that MOP-5 immobilised on support A gave good disinfection activity and resulted in the inactivation of fecal coliforms within 45 minutes of treatment exposure to sunlight. However, since the MOP-5 fixed on support A was in the powder form, its recovery from the solution for subsequent use was hindersome. Therefore, MOP-5 was immobilised onto support B.

The immobilised form of MOP-5 on support B by Friedel–Crafts reaction yielded a solid material that was easy to handle and could be filtered out for possible reuse. Moreover, it also showed a very good water disinfection activity and successfully inactivated fecal coliforms and fecal enterococci after 6 and 2 minutes of irradiation, respectively. Thus, polystyrene B proved to be a better support for immobilization of MOP-5 compared to support A, which clearly proved that the activity of an immobilized photosensitizer also depends on the type of solid support.

Thus, in the presence of solar rays, the photoactive molecules present in oils undergo photoexcitation leading to the destruction of the fecal coliforms, and the time required for SODIS could be reduced from a typical 6 hours to only one hour. The active ingredient when immobilized on a solid support resulted in increased disinfection efficiency. Although the researchers could reuse the immobilised MOP-5 without any significant reduction in photosensitizing activity, further investigation is required to ascertain the number of times the immobilised photosensitizer can be reused for water treatment. Additionally, newer plant extracts that are easy to access in the developing nations, various effective cheaper and naturally available support materials for immobilization and most importantly the economics of the process need to be studied. Such research studies clearly bring out the possibility of enhancing SODIS in natural ways by using plant extracts that are very attractive water treatment options for developing countries owing to the ease of availability of plants and plentiful sunlight, especially if the target location is a tropical area.

### *2.2.1.5 Augmenting SODIS with Bottle/Reactor Modification*

Solar collectors and reflectors are commonly used to enhance SODIS in different flow reactors. Solar disinfection improved by using solar collectors is called a Solar collector disinfection (SOCO-DIS) system. This system has proved particularly useful in inactivating the *C. parvum* oocysts. It has been reported that the greatest impact on oocyst survival was the intensity of radiation. Levels $\geq$600 W m$^{-2}$ and times of exposure between 8 and 12 h were required to reduce the oocyst infectivity in water samples with different degrees of turbidity. Other techniques of accelerating SODIS include the use of aluminum foils, lenses and mirrors in association with sunlight. Such a method was evaluated as sterilization media for the destruction of Total Heterotrophic Bacteria (THBC) using raw drinking water samples obtained from local wells in northern Jordan. The disinfection process was effective and could keep treated water for more than seven days without any further contamination.

Most scientists report that aluminum foil attached to the back of the bottles increased disinfection rate constants by a factor of 2. Some studies showed that reflective solar boxes can reduce disinfection times to 3–4 h, which also makes them very useful. Similarly, a Compound Parabolic Collector (CPC) mirror was reported to enhance the SODIS efficacy and reduced the total treatment time to disinfect the water. In another interesting modification of solar disinfection, a polyvinyl chloride (PVC) based circuit covered by an acrylic layer that was transparent to the solar rays was used for water disinfection in remote regions of developing countries. Such a system worked without a catalyst for solar disinfection of a contaminated river in Peru. Interestingly, 4 log reductions in total coliforms were obtained despite the high turbidity of the water.

Some investigations to avoid the constant requirement of PET bottles have led scientists to use a new SODIS enhanced batch reactor (EBR) design fitted with a compound parabolic collector (CPC). This was employed for the disinfection of 25 L of water in ≥6 h of strong sunlight. An important feature of this reactor was that flow was not employed as previous studies indicated that increasing flow rate lowered microbial inactivation rates. This was attributed to the fact that when microorganisms were exposed to minimum lethal UV-A doses for a short duration, rather than receive repeated doses over a long period of time, the inactivation was more effective. Needless to state, recirculated flow in a reactor would give rise to the latter condition thus hampering the effectiveness of the process. Added maintenance and circulation costs were also other factors that limit flow through systems.

### *2.2.1.6 Scaling Up SODIS*

Considerable efforts are underway to employ SODIS on a large scale to cater to a larger population in developing countries. As discussed in the preceding section, SODIS works better when the water is exposed for a short duration with a minimum lethal UV dose rather than repeatedly exposing the contaminated water for a longer time. This becomes a major limitation when scaling up SODIS using larger batch volumes or continuous flow recirculation reactors. For instance, some scientists have reported the inactivation patterns of *E. coli* during SODIS under varied UV irradiances and accumulated doses.

It has been observed that complete inactivation of *E. coli* was possible if an uninterrupted UV dose greater than a threshold limit (>108 kJ $m^{-2}$) was achieved, pointing out that bacterial killing was dependent on the UV dose rather than the UV irradiance. If the operational parameters are set such that the microbial pathogens are repeatedly exposed to sub-lethal doses of solar radiation followed by a period within which the cells have an opportunity to recover or repair, complete inactivation may not be achieved. This also has important consequences for those endeavoring to scale-up SODIS using pumped, re-circulatory, continuous flow reactors.

Moreover, SODIS application may still be possible in the (pre-)treatment of grey water despite the concerns regarding the efficiency of inactivation during recirculation and high flow rates. Some workers have reported 2 log reductions in *E. coli* levels using a 51 L continuous pilot-scale reactor prototype using grey water. Yet another scale up initiative was studied using a SODIS reactor consisting of a 25 L borosilicate glass tube fitted with a compound parabolic collector, and this was assessed under equatorial weather conditions in sub-Saharan conditions in Kampala, Uganda. Successful inactivation of *E. coli* was achieved thereby proving that the scale up of the reactor was suitable for the developing sub Saharan regions.

In another initiative, a single pass continuous flow SODIS reactor was used to treat contaminated surface water in a Kenyan rural village. This reactor used a CPC reflector to augment SODIS and successfully reduced coliform populations from 102 to 0 CFU $ml^{-1}$ after a short duration of 20 mins in a single pass; and it could deliver safe drinking water at the rate of 10 L $min^{-1}$.

In similar trials of up scaling SODIS in Spain, a sequential batch photoreactor was reported that incorporated a CPC reflector that reduced the residence time needed for disinfection. The authors reported that a higher volume of polluted water could be treated in the same time. This system also had an electronic UV_A sensor, which controlled the discharge of treated water into a clean tank following receipt of a pre-defined UV_A dose. A larger volume of the same system was estimated to produce at least around 90 L of potable water per day. Such attempts worldwide clearly point out the possibly of using SODIS for larger communities in the rural sector. Some examples of the impact of SODIS in developing countries are described in Table 2.1.

### *2.2.1.7 Economics of SODIS and Its Enhancement Techniques*

For any water disinfection technique, cost is of paramount importance apart from its effectiveness in delivering safe potable water. Thus, SODIS may not be the universal solution for safe drinking water as there are other well established methodologies such as chlorination, filtration and other household water treatment methods (HWTS) that may be as effective as SODIS. However, all these methods are always more expensive as compared to SODIS.

Therefore, SODIS is the method of choice in developing countries with very low economic standards. In a cost analysis study, it was assessed that at \$0.63 per person per year, solar disinfection was the cheapest household based intervention against waterborne disease when compared with chlorination (\$0.66), filtration (\$3.03) or flocculation/disinfection (\$4.95). In fact, over the last decade, scientists have concluded that a common aim for a household continuing to use SODIS has been based on economic benefit first rather than improvement of health and wellbeing.

Thus, it has been observed that SODIS is often the method of choice when resources are not available to employ other treatment methods. The crucial reason for this is because no operational expenses are incurred once the PET bottles are procured. It goes without much elaboration then that the SODIS enhancement technologies described in the previous sections must be assessed not only based on effectiveness but also based on their cost. The very fact that a household selects SODIS as a treatment method means that the economic conditions of the family are poor.

**Table 2.1** Examples of the impact of SODIS in developing countries.

| S. no. | Developing country/region | Impact of SODIS | Reference |
|---|---|---|---|
| 1. | Kenya–Tanzania border | • Reduced occurrence of diarrhoea in children under the age of 5 years.<br>• SODIS protected against cholera. | 6 |
| 2. | Tamil Nadu, India | • The risk of diarrhoea was reduced by 40%. | 7 |
| 3. | India: Gorakhpur (Indo-Gangetic plain), Jodhpur (arid) and Alappuzha (Keralan coastal region) | • Researchers reported a noteworthy reduction in the numbers of cases of diarrhoea/gastro-enteritis in all field locations investigated. | 8 |
| 4. | Peri-urban and rural communities in Kenya, Zimbabwe and S. Africa | • For children under age 5 years, proper use of SODIS (compliance >75%) resulted in significant reductions in dysentery incident rate ratios (incident rate in SODIS group divided by incident rate in control group) ranging between 30% and 55% in Kenya, S. Africa and Cambodia.<br>• Anthropometric benefit was seen, *i.e.* median weight-for-age was higher in those using SODIS, corresponding to a 0.23 kg difference in weight over the same period. | 5 |
| 5. | Spain, Bahrain and India | • Researchers assessed the efficacy of large volume (19 litres) transparent plastic (polycarbonate) water cooler/dispenser containers (WDCs) as SODIS reactors.<br>• SODIS was used to inactivate *Escherichia coli* and *Enterococcus faecalis.*<br>• Reduction values of 6 log10 units were obtained in all the locations tested. | 9 |

Therefore, methods used to augment SODIS that often require additives such as chemicals, titanium dioxide *etc.* are bound to be more expensive than conventional SODIS. Hence, it can be concluded that such enhancements can be feasible only when the quality of water produced is of extremely good quality (no contaminants), the speed of the process is high thereby taking less time or the volume of water treated is very huge. Only under such circumstances can the additional cost incurred be off-set.

Thus, it can be concluded that large scale SODIS reactors are economically viable when they are employed for larger communities like a school or for several households put together and efforts are underway globally to develop SODIS technologies that are economically viable.

### 2.2.2 Chlorine Disinfection

The use of chemicals especially chlorine for disinfection of water is an ancient and known practice all over the world. Chlorination is mainly used to destroy the pathogenic microorganisms in water and a secondary benefit of chlorination is odour removal to improve the aesthetic quality of drinking water. According to the World Health Organization, the adoption of drinking water chlorination has been one of the most significant advances in public health protection. Moreover, chlorine-based disinfectants are the only disinfectants that provide lasting residual protection to protect the water from waterborne disease throughout the distribution network from the treatment plant to the end user. It is important to acknowledge that as chlorination was adopted worldwide, human death rates due to diseases such as cholera and hepatitis A reduced significantly.

Chlorine is a very potent germicide as it inactivates many disease-causing microbes in water to safe concentration levels that are well within the limits of water standards. It also reduces disagreeable tastes and odors in water as it effectively oxidizes several naturally occurring substances such as sulfides and odors from decaying vegetation. Chlorine is also useful in the removal of iron and manganese from raw water by precipitation. An important and useful effect of chlorination is biological growth control in distribution systems. This is attributed to the removal of slime bacteria, molds and algae that commonly grow in water supply reservoirs, and on the walls of the piping systems and storage tanks. Thus, a residual chlorine concentration during water treatment ensures that there is no regrowth of microorganisms in the water distribution systems and enables the safe supply of drinking water to the consumer.

The mechanism of action of chlorine as a disinfectant has been extensively studied for decades as it is a traditionally used water disinfectant globally. Over the years, many theories have been proposed for the inactivation of microorganisms. Among them, some of the chief ways by which chlorine disinfection occurs are by damage to the cell surface and the nucleus, leakage of cytoplasmic material and inhibition of certain biochemical activities that are associated with the bacterial cell membrane, oxygen uptake and oxidative phosphorylation. Additionally, chlorine also affects the nucleic acids and physically damages DNA.[2]

Several bacteria and viruses were found to be easy targets of chlorination over a wide range of pH. Protozoa such as *Giardia* and *Cryptosporidium* are found to be exceptions to this rule and are much more resistant than most of the viruses and bacteria. It is interesting to note that many studies have shown complete removal of bacterial pathogens in treated water in developing countries and this resulted in reductions in diarrheal disease incidence in users ranging from 22–84%.

Water treatment systems in most developing countries use some type of chlorine-based process, either alone or in combination with other disinfectants. Chlorine can be applied in several forms such as elemental chlorine (chlorine gas), sodium hypochlorite solution (bleach) and dry calcium hypochlorite. Most commonly, sodium hypochlorite solution is used as it is less hazardous and easier to handle than elemental chlorine. Moreover, fewer training requirements and regulations as compared to those for elemental chlorine make it an attractive choice in developing countries. For instance, the treatment method for Safe Water Systems (SWSs) is point-of-use chlorination by consumers with a locally-manufactured dilute sodium hypochlorite (chlorine bleach) solution.

The Safe Water System (SWS) was developed in the 1990's in response to the cholera epidemic in South America by the Centers for Disease Control and Prevention (CDC) and the Pan American Health Organization (PAHO). The Safe Water System has been implemented in over 35 countries.

Although chlorine continues to be the ideal choice as a water disinfectant as it is relatively cheap, easy to use and effective as well, it reacts with the naturally occurring humic and fulvic acids in water to produce Trihalomethanes (THMs), which are toxic in nature. Water containing THMs, if consumed, can have a lethal effect owing to their carcinogenicity. Moreover, there are many challenges when chlorine is used alone for water treatment, which include treating resistant pathogens such as *Giardia* and *Cryptosporidium*, new environmental and safety regulations regarding residual chlorination and strengthening operational safety at the treatment facilities during the handling of chlorine chemicals.

Chlorination is most appropriate in areas with a consistent supply chain for hypochlorite solution with relatively lower turbidity water, and in emergency situations where educational messages can reach users to encourage correct and consistent use of the hypochlorite solution. The cost of chlorine is also a major factor that will impact its use especially in developing countries that have a low income. Typically, a bottle of hypochlorite solution that treats 1000 liters of water costs about 10 US cents using refillable bottles and 11–50 US cents using disposable bottles, for a cost of 0.01–0.05 cents per liter of treated water. While chlorination remains the most commonly used disinfection method by far, its use may be limited due to the potential health hazards associated with the formation of disinfection by-products such as THMs.

### 2.2.3 Filtration

Filtration is one of the oldest methods of water purification that has been employed since ancient times. As it is an easy technique and relatively inexpensive, it has found widespread use in many parts of the world especially in developing countries. The main reason for the widespread use of this simple method may be because it relies on the basic law of nature that is gravity and for eons, humankind has observed this process in daily life, for instance the flow of water through a bed of sand.

Over centuries, this simple technique has seen several changes and modifications mainly to bring about enhanced effects. In the context of water purification, one of the oldest techniques of filtration that has been reported is slow sand filtration. Since sand is easily available in almost all parts of the world, it is natural that it was one of the initial materials used to filter contaminated water. However, the process of filtration took a long time and after several uses, clogging of the filter owing to the deposition of organic and inorganic material as well as the biofilm that formed due to the growth microbes on the filter surface resulted in decreased efficiency of the filter. This led to further research to alleviate these issues that resulted in the use of other filter materials such as activated carbon, ceramics and a host of natural materials such as rice husk *etc.*

Apart from this, improvement of the filter material by coating it with antimicrobial substances such as silver was yet another approach to enhance the process of filtration. Technical improvements such as pressure driven filtration, backwashing to rectify clogged filters and increase their life span, are other areas that have been investigated. In developing countries, household water filters have been especially studied for their simplicity and economic feasibility. With increasing environmental protection over the decades, recycling has received attention and there have been many attempts to study the recyclability of filter cartridges so that the plastic body as well as the filter material is regenerated and reused. Yet another important aspect of filtration is the myriad factors involved in its successful implementation.

Factors such as quality of water and the level and type of pollutants (physical, chemical and biological), source of water (groundwater or surface water), type and quality and composition of filter material, durability of the filter, efficiency of filtration, availability and cost play a crucial role in the successful implementation of filtration in any developing nation. Considering the above changes and novel initiatives that have be taken in the arena of procuring safe potable water by filtration, particularly for low economic group countries, it is imperative that each of these attempts is suitably highlighted. The following section therefore discusses this in detail.

#### 2.2.3.1 Slow Sand Filtration

This type of filter is very simple to design and it essentially consists of a bed of sand, which is normally around 24 inches thick. This layer is topped with a layer of gravel into which a drain pipe is embedded, which allows the system to be drained after use. The water to be treated is passed through these layers and over time, this results in the formation of a biological layer called 'schmutzdecke'. Thus, filtration of the polluted water occurs not only due to the physical process but also due to the action of the thin biological layer that forms on the filter media. This process effectively removes most of the turbidity, organic and inorganic contaminants, bacteria and some protozoan oocysts such as *Giardia* and *Cryptosporidium*.

Although no pre-treatment is needed for filtration, high turbidity or the presence of algae in water may necessitate simple retreatment such as sedimentation or flocculation to ensure that the raw water entering the slow sand filter is fit for filtration and does not clog it. With repeated use, when the schmutzdecke layer thickens, some of it is scraped out to expose a fresh layer. This improves the filtration rate. Nevertheless, after a few months, as the sand layer decreases in depth, new or recycled sand is added again to continue the filtration process. Factors such as the quality and uniformity of the sand and gravel particles affect the filtration efficiency. Generally, the filtered water does not require any further treatment but chlorination is routinely carried out to ensure the safety of the filtered water and to prevent recontamination.

Literature reports have described the use of slow sand filtration to remove a spectrum of microorganisms from various source waters in developing countries across the globe. Most of the studies indicate that slow sand filtration typically removes around 97% *E. coli*, 99% *Giardia*, *Cryptosporidium*, helminths and protozoa, 95% iron and almost 90% arsenic. However, it must be noted that some pollutants such as magnesium and calcium are not effectively removed by these filters. Moreover, complete elimination of all pathogen bacteria and viruses cannot be guaranteed and this makes disinfection by other means such as chemicals post filtration mandatory. It is interesting to note that slow sand filtration has been explored as an alternative for large scale river bank filtration and it has proven to be extremely effective in purifying water contaminated with high levels of fecal coliforms.[14]

### *2.2.3.2 Ceramic Filters*

Filtration by means of ceramic filters is a simple and effective technology that can be used in developing countries and they are often denoted as CWFs (ceramic water filters). One of the main reasons for this is that the filters can be produced by the local populations by utilizing materials and techniques that do not rely on any external source. Moreover, these filters are environmentally compatible and can successfully be used equally to treat unpurified and/or inadequately disinfected water. Due to its worldwide acceptance and use during the last few decades, several NGOs and various stakeholders have shown interest in promoting this technology at the grassroots level leading to augmented efforts in the production of ceramic filters.

For instance, 'Potters for Peace' is an organization that is now established with over 33 production facilities in developing countries that facilitate in promoting and supplying ceramic filters to house-holds. It is fascinating to note that one study conducted in the year 2012 reported that more than 4 million people all over the world use more than 700 000 ceramic pot filters to treat water for their daily potable water requirements. Like any other water treatment method, ceramic filters are also equally efficient in removing a plethora of microorganisms such as Total coliforms, *E. coli*, fecal coliforms and a host of protozoa and some viruses. It has been shown that these high elimination rates of bacteria especially coliforms are very useful in bringing down the incidences of Early Childhood Diarrhea, especially in developing countries.

Scientists have reported 99% reduction rates for *E. coli* across several laboratories and field trials have indicated ceramic filters as 'Protective', as assigned by the WHO, for bacteria. Investigations using bacteriophages such as MS2 have revealed that ceramic filters are effective in removing viruses but the efficiency of the filters was enhanced when they were coated with silver. For instance, in one such study, it was observed that the virus removal capacity for MS2 was 0.6 to 0.9 after 5 weeks and this increased to 1.1 to 1.8 after 13 weeks with silver impregnated filters. Similar observations have been made for *E. coli* as well but in this case, the contact time with silver was the determining factor.[15]

Other coating materials such as a polymer based quaternary amine functionalized silsesquioxane called 'TPA' have also been shown to be a viable alternative to silver due to their good antimicrobial properties and lower cost compared to silver. This would be an advantage for the implementation of this technique in developing nations. However, one concerning factor could be the release of TPA in the effluent and its impact on human health that need to be investigated thoroughly.

Researchers have also found that human behaviors such as compliance and declining microbial effectiveness due to improper maintenance are very important factors that determine the success rate of such interventions. Thus, the ability of the filter to perform decreases if it is not maintained well. Therefore, scientists recommend cleaning these filters at least once every 4 months. Yet another factor that is of paramount importance are the guidelines followed by the local authorities. It is alarming to note that many developing countries do not have a standard for protozoan and viruses in place. Mostly, the water after passing through the filtration process is only checked if it is bacteria free. Therefore, researchers in this field have emphasized the need to draft new national policy for evaluation of water filters by regulating authorities that can be used as a production standard.

According to the United States Environmental Protection Agency (USEPA), standards for testing microbial purifiers recommend a 6-log reduction of bacteria, 4-log reduction of viruses and 3-log reduction of protozoan cysts. Overall, ceramic water filters are an effective method for drinking water treatment in developing countries but the main challenges are sustainability, community compliance and maintenance of the filter. Nevertheless, research in these areas can surely strengthen the deficient areas and augment their application globally in the years to come and many such examples are discussed in the section on Hybrid Filtration Methods.

### 2.2.3.3 Rapid Sand Filtration

Another important and major variant of filtration technology that has been commonly used for water treatment for centuries is rapid sand filtration. Due to its simplicity and wide acceptance as well as an established methodology to obtain safe and pure drinking water, rapid sand filtration is still in use globally, especially in developing countries. Rapid sand filtration can be used to remove physical, chemical and biological contaminants as well. This type of filter is also called a rapid gravity filter and it is very commonly employed as a part of the water treatment process in most municipal drinking water treatment facilities. The use of this method is decades old and it has been widely utilized in large municipal water systems mainly because rapid sand filters require smaller land areas as compared to slow sand filters.

These filters use coarse sand and other granular media to remove pollutants. Usually, this is preceded by flocculation and sedimentation and followed by a disinfection step. Rapid sand filters either operate under gravity or pressure. Maintenance as with any other filter is an integral requirement, which could be several times a day and is usually carried out by backwashing or passing compressed air. The literature includes several studies where this technique has been successfully employed.[16] For instance, a commonly used methodology to remove iron and manganese from groundwater in the Netherlands is aeration followed by rapid sand filtration.

The mechanism of removal is attributed to three processes that occur in the sand filter, *viz.* homogeneous oxidation, heterogeneous oxidation and biological oxidation. Similarly, there are reports of a spectrum of microbes that can be easily and effectively eliminated by using rapid sand filtration including some pathogenic microorganisms. However, virus removal efficiencies are difficult to monitor duc to the technical challenges in quantifying low viral concentrations present in drinking water.

Fascinatingly, over the last decade, due to positive improvements in microbial assessment methodologies, this problem has also been overcome to a large extent. This was demonstrated in an interesting study conducted in Bangkok, Thailand[16] where the removal efficiencies of viruses were monitored using real time polymerase chain reaction (qPCR). Water samples containing pepper mild mottle virus (PMMoV) and JC polyomavirus (JCPyV) were treated with rapid sand filtration and coagulation-sedimentation (CS) as individual processes and compared. The researchers observed some seasonal variation, *i.e.* during the wet season, PMMoV was eliminated more by rapid sand filtration as compared to CS while the reverse was true for JCPyV. Interestingly, during the dry season, both the viruses were removed similarly with CS being more effective. Such studies point out the potential of rapid sand filters in removing pathogenic enteric viruses with the added information on specific environmental conditions that are suitable for enhancing the performance of the filters, in this case, the wet season.

### 2.2.3.4 Activated Carbon Filtration

Activated carbon filtration is a very common and popular method to remove contaminants in water due to its inherent ability to absorb impurities from any aqueous solution. Therefore, this technique has played a pivotal role in water purification since ancient times, especially in the developing world, where it is used both at individual household and community levels equally. It is also commonly used as a part of the entire water treatment scheme in many parts of the world and as a pre-treatment method prior to reverse osmosis and ultraviolet radiation. Activated carbon has a remarkable ability to trap myriad chemical, physical and biological pollutants commonly found in drinking water.

This ability is due to the chemical structure and surface properties of activated carbon. Activated carbon is processed carbon with a slightly positive charge added to it so that it can remove impurities. Moreover, it is highly porous and therefore has a high surface area to volume ratio, which in turn increases the rate of absorption. Carbon is easily obtained from various sources like coconut shell, wood or coal and it is subjected to an activation process either by steam processing or chemical methods. However, due to ongoing global deforestation issues, wood may not be a viable choice and this has triggered some research to find other alternatives like locally sourced agricultural byproducts, especially in developing countries.

Granulated activated carbon is more commonly used than activated carbon blocks as the latter cannot be reused and needs frequent replacement. It is interesting to note the type of investigations reported in the literature for optimizing filtration by activated carbon filters. One such attempt has been to enhance the surface properties of the filter as the application of activated carbon mainly depends on the surface chemistry and pore structure of the porous carbons, as discussed earlier. The method of activation and the nature of precursor used greatly influence the surface functional groups, which in turn determines the type and quantity of impurities that can be absorbed on the activated carbon.

Recently, a plethora of techniques such as acid or base treatments, ozone treatment, surface impregnation, plasma and microwave treatment have been investigated to develop surface modified activated carbons for use in water purification.[17] Another important aspect studied by researchers globally is the ability of activated carbon to eliminated new and emerging contaminants that are often encountered currently. One such pollutant is the poly and perfluoroalkyl substances (PFASs) that have been detected in several potable waters at high concentrations worldwide and have also led to the implementation of regulatory guidelines in many countries such as Europe and the US.

It is fascinating to note that scientists have found granulated activated carbon filters to be very effective in removing PFASs and the linear isomers had greater removal efficiencies than the branched ones. Yet another interesting study reported the removal of microcystin toxin (a cyanobacterial toxin) from drinking water by using either granulated carbon filters or activated carbon filters. Thus, activated carbon filtration continues to be a method of choice in several parts of the world and considerable efforts to enhance this process are underway as during activated carbon filtration, chemical and biological impurities get adsorbed on the surface of the filter thereby reducing the filtration capacity and efficiency over a period of time. Moreover, the adsorbed microbial impurities may start colonizing on the carbon surface and water passing through such a filter may get contaminated rather than getting purified.

#### 2.2.3.5 *Natural Filters*

Any water treatment technique that is designed for developing countries should be efficient in removing impurities and must be cost effective. Most of the filtration methods discussed in the previous sections meet this requirement. However, increasing mortality rates due to incidents of water borne diseases have triggered some interest in exploring the possibility of employing economical and naturally available materials as filters in low-income countries. An innovative study was conducted with wood as a filter element for water treatment where the investigators examined the possibility of filtering water in the perpendicular direction of the wood fibers.[18] This prompted some interest to investigate the possibility of using wood for water filtration and further studies were conducted.

Pilot systems were constructed using three different species of wood, *viz.* caixeta, garapuvu and pine, and two modes of filtration, *i.e.* dead end and helical cross flow. The study showed that porosity, wood density and the mode of filtration employed were critical factors that affected the efficiency of water filtration. The more porous the wood, the greater the filterability. It was also observed that a wood density of 0.5 $g\,cm^{-3}$ and porosity of 40% proved to be noteworthy factors that indicated that there was a correlation between both these factors that had to be considered while choosing the type of wood for water filtration. Moreover, among the three types of woods studied, pine was found to be superior and helical cross flow filtration was much better than the dead-end mode, which could be further enhanced by employing coagulation as a pre-treatment. In general, the authors state that water filtration using wood has potential to be used as a cost-effective methodology with negligible power consumption as the working pressures required in such systems are very low.[18]

Efforts along the same lines have led to the innovative use of plant xylem from the sapwood of coniferous trees for water filtration. In a pioneering study, scientists taking cue from nature used plant xylem as a water filter to remove pollutants. Xylem is a porous material that conducts fluids in plants from their roots to their shoots and the pore size of xylem is ideal (typically a few nanometers to a maximum of 500 nm) for filtration to remove pathogenic microorganisms. Therefore, the investigators prepared a filter by removing the bark of pine tree branches and inserting the xylem tissue into a tube.

The study demonstrated that xylem fibers could effectively remove bacteria up to 99.9% with flow rates of around 4 L per day through 1 $cm^2$ filter areas at an applied pressure of approximately 5 psi. This was found to meet the drinking water needs of one person. Thus, a simple method of using naturally available plant xylem fibers that are biodegradable and cost effective proved to be a potential technique to address the potable water needs in developing countries. Further research along the same lines is required so that this method can be applied on a bigger scale to meet the safe water requirements of a larger population.[19]

### 2.2.3.6 *Hybrid Filtration Methods*

Over several decades, a plethora of filtration approaches have been tried and tested and many of them were successful in eradicating chemical, physical and biological impurities that are usually present in water. Many such techniques are discussed in the previous sections. However, it must be understood that there is no universal filtration method that can ensure the complete removal of all pollutants. Moreover, factors such as source water, environmental factors, geographical location and most significantly the efficiency of the technique along with its feasibility and economics determine the preference of one technique over another. This is especially important and of paramount significance in developing countries. Consequently, scientists all over the world have tried to use a combination of methods to come up with an effective and affordable filtration procedure. Here, we refer to this as 'Hybrid Filtration'. Some examples of this are listed in Table 2.2.

It is evident from the examples discussed that a number of myriad variants of filtration that result in 'hybrid filtration methods' are possible, and each combination has its own pros and cons. Most of the combination methods use either different types of filter materials or adsorb metals such as copper or silver. Metals are well known for their germicidal action and therefore some hybrid techniques also involve filtration followed by storage in metal coated or metal pots. Contact with a metal surface for some time will result in the inactivation of many pathogenic microorganisms and subsequently reduce the incidences of waterborne illnesses that challenge the global population.

Yet another remarkable variation appears to be the use of water treatment trains that consist of a series of steps (water treatment methods) that finally deliver safe potable water to the end user. Though this is a very good option as each treatment step in the train usually compliments the next, many steps in the treatment train can lead to an increase in cost and take a long duration. However, several attempts are being made to reduce the number of steps and innovate by optimizing and enhancing the existing methods. For instance, a combination of red clay, sand, and wood sawdust has been used successfully as a filtration system called 'Terafil' in Orrisa, India.

**Table 2.2** Examples of hybrid filtration methods suitable for use in developing countries.

| S. no. | Hybrid filtration method | Mechanism of action | Developing country | Contamination removed | Reference |
|---|---|---|---|---|---|
| 1. | Slow sand filtration and copper | Water filtered through SSF is stored in copper containers where copper acts as a disinfectant | Highly recommended for developing countries | Bacteria and Viruses | 16, 20 |
| 2. | New generation aquaguard POU | Consists of a combination of a candle prefilter, activated Carbon filter and UV irradiation | South Africa | Viruses, Bacteria, Phages, *Cryptosporidium* oocysts | 16, 20 |
| 3. | POU carbon filters and copper/silver | Water passed through a hybrid filter containing carbon + 10% copper alloy + 0.1% silver | Tap water studied in the US, recommended for use in developing nations | *Legionella pneumophila* | 17, 21 |
| 4. | PAC-MF (powdered activated carbon-microfiltration) | Water passed through a reactor consisting of a PAC zone and a MF filtration zone. | River water laden with sewerage effluents, Tokyo, Japan | Organic matter and viruses | 18, 22 |
| 5. | Filtration and metals | Raw water was filtered through a clay pot filter made of terracotta clay and sawdust followed by passage through a copper wire mesh | Lobamba and Mtilane River water, Swaziland, South Africa | *E. coli*, total coliform, total hardness, turbidity, electrical conductivity, cations and anions | 16, 20 |

| | | | | | |
|---|---|---|---|---|---|
| 6. | Coagulation, ozonation, ceramic UF, GAC filtration and disinfection | Pre-treatment by coagulation, flocs and microbes were removed by ceramic UF and oxidation of organic pollutants and to desorb the flocs, ozonation was set up and finally GAC removed the residual organic matter, ammonia, EDCs, PPCPs, *etc.* | Micropolluted surface water from Dongjiang River and the city canal, southern China | Coliforms, turbidity, dissolved organic matter, ammonia *etc.* | 19, 23 |
| 7. | Terafil | Filter produced by combination of red clay, sand and wood sawdust | Orrisa, India | Microorganisms, sediments, dissolved iron, colour, odour, heavy metals | 16, 20 |
| 8. | Tata Swach | POU water filter made of rise husk ash functionalised with nanosilver | India | Bacteria, viruses, cysts | 16, 20 |
| 9. | Filter made of iron coated sand and charcoal fitted to a hand pump | Filter made of iron coated sand and charcoal adsorbed the arsenic and almost 70–80% arsenic was removed from the drinking water. The filter included an inlet nozzle that was attached to the outlet of a hand pump. | West Bengal and Chhattisgarh districts, India | Arsenic | |

Terafil has been reported to efficiently remove physical (sediments), chemical (heavy metals) and biological pollutants (6-log reduction in *E. coli* and 4-log reduction in various viruses) from water due to its non-interconnected pores mainly because of the presence of clay membranes. Moreover, clogging of filters could be prevented due to this feature. To further improve filtration performance, lightly scraping the surface of the Terafil disc is recommended, thus exposing a fresh filtration area. Flow rates of 2 $L\,h^{-1}$ to 1500 $L\,h^{-1}$ could be achieved and a life span of five years is estimated for a disc. It is most suitable for treating water in developing countries due to its low cost and ability to treat both ground and surface water.[20]

Yet another interesting initiative along similar lines consists of a water filter attached to hand-pumps, which can remove arsenic from drinking water. The inlet nozzle of the filter is designed to fit into the outlet of a hand pump. The filter material essentially consists of iron coated sand and charcoal. Initial pilot studies conducted in West Bengal and the Chhattisgarh districts of India reported almost 70–80% arsenic removal from drinking water due to the high adsorption affinity of arsenic towards iron oxide. This filter is ideal for developing countries due to its uncomplicated design and operation. Water to be purified initially passes through a micron filter (muslin cloth) that removes debris and dirt and subsequently through beds of adsorbents (iron oxide and charcoal) that adsorb the arsenic present in the water. The filter is assembled in a simple and lightweight PVC pipe. Moreover, the use of a hand pump saves energy and makes it an economical option for the rural population. The proposed cost is around INR 2000 with a capacity to purify an average of 20 000 liters of water.

### *2.2.3.7 Economical Aspects of Filtration Techniques*

As these techniques are specifically discussed with respect to developing nations, it is of paramount importance that the filtration methods are available to the population at inexpensive rates. Moreover, the materials and set-ups required should be easily available and operation and maintenance should be nominal so that skilled labor is not essential. Most of the methods described in the previous sections meet the criteria and are therefore suitable for low income countries. In fact, many of these methods have been and are being used successfully in developing countries worldwide. Construction costs are an important factor that must be considered while setting up a filtration unit.

For instance, in the case of slow sand filtration, large expanses of land are needed but the input of construction material is very low and hence the capital costs only consist of labor wages and costs of land procurement. Operation and maintenance costs are mostly incurred for cleaning the filter bed. Since there is no requirement of backwashing in terms of compressed air, mechanical stirring or high-pressure water, there are huge savings on fuel and electricity or utilities costs too. On the other hand, the construction cost of rapid sand filters is determined mainly by the cost of materials such as cement, building sand, pipes and valves, and additionally, the cost of land and transport of materials could augment the expense significantly.

Regarding the operation and maintenance cost, the energy required to operate a rapid sand filter as well as the cost of backwashing is huge. Moreover, skilled labour and trained personnel may be needed to monitor the process. Thus, in general, the cost of rapid sand filters is higher as compared to that of slow sand filters.[14] A literature study reported that a typical ceramic filter costs US$3.50 per unit and the filter may cost anywhere between US$0.49 and US$1.02. This is for a scale of operation ranging between 1 to 11 $L\,h^{-1}$. Other hybrid filters such as Terafil[20] can generate typically 2 to 1500 $L\,h^{-1}$ and cost around US$0.5 for the filter and approximately US$4 for the entire unit. Similarly, Tata Swach and its multiple variations have a running cost of only US$4 and a capital cost of US$50 generating about 3000 L of treated water. Thus, all these techniques are very suitable for developing countries where electricity is not available and are mainly useful to eliminate turbidity, microbial contamination and metals (specific for arsenic, fluoride *etc.*) in many cases [20].

### 2.2.4 Herbal Approaches

Among the various economical techniques available for treating water in developing countries, natural compounds such as herbs appear to be a promising method. Herbs are typically plant materials that are used to treat drinking water and any part of the plant can be used depending on the species and the location of the active ingredient in it. For instance, seeds, leaf, roots, stems and flowers of a plethora of plant species are known to have medicinal properties such as anti-microbial action, coagulating properties and even virus inactivation ability. Moreover, they are equally effective in removing turbidity, colour and odor from water thereby making it aesthetically pleasing. The practice of using herbal substances to treat drinking water is very ancient and has been applied in many parts of the globe. However, this technique has not been considered as a potential and promising way to obtain safe potable water for the underdeveloped world.

For instance, 'Ayurveda' is the Indian system of natural healing and is based on ancient texts such as the Charaka Samhita and Shusruta Samhita, which can be traced back to almost 300AD. These ancient texts are store houses of important and practical information that states different ways of purifying water such as the use of wood of Ayla (*Planthouse emplace*) and Tuli (*Oscimum sanctum*). These substances exhibited antibacterial and insecticidal properties and were very good clarifying agents. Similarly, in another part of the world, Sudan, North Africa, the drumstick tree (*Moringa oliefera*) is popularly known as the 'clarifier tree' and has been used to treat water for decades. The seeds from this tree have been reported to be one of the most effective primary coagulants for water treatment especially in rural communities.

**Table 2.3** Examples of some herbs for water treatment in developing countries.

| S. no. | Herb | Effect on pollutants | Developing country | Reference |
|---|---|---|---|---|
| 1. | *Moringa oleifera* | 90.99% reduction in fecal coliforms | Ghana, West Africa | 20, 24 |
| 2. | *Moringa oleifera* + Alum | Effective coagulant when used with alum. Turbidity reduced beyond levels found when alum was used alone. | Sudan, North Africa | 21, 25 |
| 3. | *Moringa oleifera* | Doses of 30 to 55 mg $L^{-1}$ *Moringa oleifera* stock solution needed for turbidity values between 40 and 200 NTU | Malawi, East Africa | 22, 26 |
| 4. | *Moringa oleifera* + Sudanese bentonite clays | Used to treat Nile river water. 90% reduction in *S. mansoni* that causes schistosomiasis | Sudanese villages, Africa | 23, 27 |
| 5. | Tulsi leaves (*Ocimum sanctum*) & Neem leaves (*Azadirachta indica*) | Neem and tulsi alcoholic extracts were found to be more effective in reducing the MPN of all tested water samples and inhibiting *Salmonella typhi* | Pune & Satara district, Maharashtra, India | 24, 28 |
| 6. | Tulsi (*Ocimum sanctum*) | 500 mg of $l^{-1}$ extract treated water showed 95 to 98% antibacterial activity in 14 to 16 hours using the Plate count method of analysis. | River water from Kavery at Trichy, Tamil Nadu, India | 25, 29 |
| 7. | *Tamarindus indica* fruit crude pulp extract | Optimum dose of 3000 mg $L^{-1}$ resulted in highest turbidity removal efficiency of 99%. However, pH adjustment was needed to reduce acidic nature of treated water. | Kubanni Reservoir, Ahmadu Bello University (ABU) Zaria, Nigeria | 26, 30 |
| 8. | Purified okra (*Hibiscus esculentus*) seed proteins | Approximately 100% inactivation of fecal and *E. coli* count in raw water was achieved. | Bourn Brook canal, adjacent to the University of Birmingham Train station. | 27, 31 |
| 9. | Neem plant (*Azadirachta indica* A. Juss) | *Azadirachta indica* extracted oil with ethanol exhibited a microbial reduction of 99% for ground water, 98% for well water, 96% for lake water and 95% for river water at a minimum dose of 5 ml $l^{-1}$ after 12 hours of treatment time. | River, well and lake water, India | 28, 32 |

For the last few decades, scientist have started investigating this natural resource and several studies have been conducted to ascertain the effectiveness, type and number of plant species and corresponding herbal ingredient, and type of application in water treatment, *i.e.* whether herbal treatment can be employed for its viricidal and microbicidal properties and/or removal of chemical and other physical contaminants. The main reason for this shift towards herbs is their economical nature and ease of availability in developing countries as many indigenous plants grow worldwide under different topographical conditions and these can be exploited as inexpensive materials to treat water both at the household and community level. Table 2.3 lists some of the main herbs that have been used to treat water in developing countries.

#### *2.2.4.1 Moringa Oleifera*

One of the most commonly used natural substances to treat water is *Moringa oleifera* (Drumstick). The main reason for this is the abundantly growing trees in most parts of the world, which results in the ease of availability for the local population. A spectrum of research work based on this useful tree and its application for treating water is available in the literature and many scientists have reviewed it in detail. Highlights of such studies are presented in this section focusing on the relevant discussion for water purification in developing countries.

It is interesting to note that the tree is very slender with drooping branches, which are regularly cut to allow it to regrow well and maintain an easy reach of the different tree parts, all of which are useful. *Moringa oleifera* is native to various countries such as India, Africa, Arabia, South East Asia, South America and the Pacific and Caribbean Islands and it is known by common names in different parts of the world, Drumstick being the most popular. The main usage of *Moringa oleifera* seeds in water treatment has been as a coagulant and as a disinfectant. Many investigators have undertaken research to ascertain the active ingredient of the seed so that a better efficiency of treatment can be attained. Some scientists report that shelled *Moringa oleifera* contains around 36% proteins, 34% lipids and 5% carbohydrates and the unshelled *Moringa oleifera* contains a lower amount of proteins, lipids and carbohydrates.

The active ingredient has been identified as a polyelectrolyte that is responsible for its coagulating property. Several studies have been undertaken to ascertain the active principle in *Moringa oleifera* and all of them have confirmed the same. Yet another investigation on the active anti-microbial agent present in the seeds pointed out the presence of a glycosidic mustard oil, 4α-4-rhamnosyloxy-benzyl-isothiocynate. This compound was reported to be soluble in water and non-volatile in nature.

Among various applications in water treatment, softening of water with *M. oleifera* has also been studied. *M. oleifera* was observed to be extremely useful as it could successfully soften water and at the same time it was accompanied by a very low reduction in alkalinity, which is usually a prerequisite for providing buffering capacity to achieve the required treatment goals. Disinfection of water need not be emphasized further as the most important target of any water treatment scheme is inactivation of microbes. This is very crucial in developing countries as the local population are more vulnerable to drinking contaminated water due to a lack of basic knowledge as well as cross contamination of sewers with the drinking supply. Most of the waterborne illnesses ensue because of consuming pathogen laden unsafe water. Although many conventional disinfectants such as chlorine are a common choice, they may not be an appropriate option in developing nations as they may have to be imported and can be a costly choice. Moreover, disinfection byproducts of chlorine are known to be carcinogenic. In this scenario, natural herbs and plant materials are an apt choice as disinfectants and therefore *Moringa oleifera* has been studied by various scientists globally as a disinfectant.

Several reports state that the active ingredient, *4α-4-rhamnotyloxy-benzyl-sothiocynate*, isolated from defatted *M. Oleifera* seeds has antimicrobial action. A variety of bacteria covering both gram positive and gram-negative species are susceptible to the active ingredient found in the seeds of this plant. For instance, bacterial species such as *Bacillus Subtilis, Serratia Marcescens, Mycobacterium Pheli, Coliforms, E. coli*, and *Enterobacter aerogenes* and viruses such as Herpes Simplex Virus have all been shown to be inactivated by *M. oleifera* and this is indeed very promising as it covers a whole range of possible pathogens that may be encountered in raw water. Many studies report the action of *M. oleifera* on *E. coli* and it is interesting to note that kinetic models have been developed for the same, and in one such investigation, the coefficient of specific lethality for *E. coli* inactivation was reported as 3.76 L $mg^{-1}$ $min^{-1}$.

Rupturing the cell and damaging the intracellular contents appear to be the mechanism of action of *M. oleifera* seeds, as reported by most experts.[24–27] Thus, *Moringa oleifera* has been found to have several applications in water treatment and therefore such applications should be encouraged in rural areas of developing countries where the growth of the tree is abundant and can result in a viable, simple, economical and cost effective methodology for safe potable water. However, the toxic effects of using *Moringa oleifera* in water treatment cannot be overlooked. Cytotoxity and genetoxicity of powdered *Moringa oleifera* seeds in the concentration range of 1 to 50 mg $L^{-1}$ have been reported. Therefore, care must be exercised in the application of these natural coagulants in water treatment.

### 2.2.4.2 Tulsi

*Ocimum tenuiflorum* (Tulsi) is a shrub that is commonly found in South Asia and most abundantly in India and some parts of China, the Middle East,

North Africa and Australia. It is considered as a very holy plant in India and every house always has a tulsi plant at the entrance of the house. Since ancient times, Tulsi has been lauded for its medicinal properties as documented in the Ayurvedic and Unnani systems of medicine for holistic health. It has been used in households for myriad purposes such as medicinal, culinary and religious purposes. It is very common to witness tulsi added to holy water in Indian temples and it is distributed to devotees as it is believed that tulsi disinfects the water and imparts a pleasant taste to it.

There are two varieties of tulsi based on the leaf colour. Tulsi plants that have green leaves are commonly referred to as Sri or Rama Tulsi and the ones with purple leaves are called Krishna tulsi colloquially. The medicinal uses of Tulsi are manifold and it has been used to cure common ailments such as cold, cough, flu, fever, coli pain, sore throat, hepatitis, malaria and diarrhea. It is interesting to note that the leaves are supposed to be good to sharpen the memory. The active ingredients present in tulsi have been researched extensively and it has been elucidated that the leaves contain 0.7% volatile oil that consists of approximately 70% eugenol and around 20% methyl eugenol. Constituents such as phenolic compounds, flavonoids, terpenes and different sterols are also reported in fresh tulsi leaf and stem extracts.

One pivotal finding is that tulsi leaves exhibit anticancer activity and scientists have reported that the alcoholic extract of the leaves has an influence on carcinogen metabolizing enzymes such as the cytochrome P450, which are important in the detoxification of carcinogens. Other equally fascinating findings of tulsi include its radioprotective, antihypertensive and cardioprotective effects on animal models. Experiments on rats and dogs have also indicated that tulsi has analgesic, antipyretic and anti-inflammatory effects and it delayed the process of cataractogenesis in experimental models.[28] Such positive results have spurred interest among scientists to explore the antimicrobial activity of tulsi and the results are indeed very remarkable. It exhibited antimicrobial activity against a host of microbes ranging from *E. coli* to *B. anthracis*, *B. subtilis*, *Staph. aureus*, *P. vulgaris*, *Mycobacterium tuberculosis*, and *Pseudomonas*, to name a few. Moreover, water pathogens such as *Vibrio cholerae* are also inactivated by tulsi leaf extract.

An Ayurvedic preparation containing tulsi leaf extract along with other herbs has been shown to have anti-malarial action. These findings have led researchers to experiment with tulsi in water treatment. In one such study, tulsi leaf extract was added to tap and river water and the authors found that maximum activity was obtained at 600 $mg\,L^{-1}$ at pH 6.8–7.0 after a duration of 15 to 16 hours.[29] Thus, tulsi holds promise just like the other herbs discussed in this section as a disinfectant in developing countries as it is a simple, cost effective technique and treated water will not only be free from disease causing microbes but will also act as medicinal water due to the plethora of therapeutic effects of tulsi.

### 2.2.4.3 Tamarind

*Tamarindus indica*, commonly called tamarind, is a leguminous tree and is reported to be native to tropical Africa and to the Indian subcontinent. It is also found to be widely distributed in the tropical belt from Africa to South Asia, northern Australia and in Taiwan and China. Tamarind fruit pulp is frequently used for culinary purposes for its distinct sour taste. Apart from this, it has been used as a folk medicine to relieve constipation. This is mainly due to the presence of malic and tartaric acids that impart a laxative effect when consumed. The use of tamarind in water treatment appears to be largely as a coagulant to treat turbid waters.

There are reports of studies conducted during the late 1970s in Nigeria where the local people found the use of *T. indica* for water purification. Scientists have tried to elucidate the efficiency of turbidity removal by *T. indica* and in one such study on a highly turbid (478 NTU) raw water sample, it was observed that when concentrations ranging from 500 to 3000 $\text{mf L}^{-1}$ of *T. indica* were used, the turbidity decreased from 64 to 99% and the optimum dose was found to be 3000 $\text{mg l}^{-1}$. The main drawback of this method is the observed high acidity levels after treating with *T. indica*, which require pH adjustment of the treated water. Nevertheless, researchers reported its usefulness in drastically decreasing the turbidity of raw water, particularly in rural communities, as it is an economical, and socially and technically feasible application.

Yet another interesting application of *T. indica* in water treatment has been in defluorination of fluoride containing water. Fluoride removal from water has been considered by using activated and $MnO_2$ coated tamarind fruit shell in both batch and column studies and efficient fluoride removal has been documented. In a similar attempt, twelve groundwater samples from fluoride affected areas of Anathapur District, Andra Pradesh, India were treated with *T. indica* fruit shell and the adsorbent capacity was found to be significant but other anions present in the water sample also affected the rate of adsorption. Literature reports used tamarind in water treatment mainly for the removal of turbidity and chemical pollutants. Although there are several studies that report the antimicrobial properties of tamarind, there have been no specific experiments on the antimicrobial activity of *T. indica* in water treatment.

For instance, the medicinal uses of tamarind are well documented and there are reports that suggest that ethanol extracts of *T. indica* produce strong antibacterial activity against *E. coli*, *Kleabsiella pneumoniae*, *Salmonella paratyphi A* and *Pseudomonas aeroginose*.[30] Many of these bacteria are common water pathogens and therefore it may be expected that tamarind extract will be able to disinfect water by inactivating these pathogens and detailed investigations regarding its use as a water disinfectant should be undertaken. Therefore, from the preceding discussions, it appears that *T. indica* can be effectively used as a natural coagulant to reduce turbidity and chemical contaminants such as fluorides but its potential antimicrobial application needs further research to make it more useful as a natural choice for water treatment in low-income nations.

### 2.2.4.4 Neem

*Azadirachta indica* or Neem is a tree belonging to the Miliacea family that is abundantly found in the Indian subcontinent. Neem has been well known for its antimicrobial properties since ancient times and in Ayurveda, that is, the Indian system of treating diseases, Neem has been endorsed as the one medicine that can cure almost all ailments. Over decades, as people moved from one place to another, the emigrants took the neem seeds with them and planted them in other regions as it usually grows well in tropical and sub-tropical climates. Thus, the neem tree was introduced to regions such as Australia, Africa and South America. Being a medicinal plant, neem has been used by natives in these countries to relieve a plethora of ailments such as skin ulcers, intestinal diseases, and gastroenteritis.

A remarkable aspect of this tree is that all parts are tremendously useful in some way. For instance, the leaves can be used to treat diabetes, skin infections and eczema and are an antipyretic (fever reduction). Neem barks have been used since ancient times as a toothbrush due to the valuable effects they have on the teeth and gums. The roots are also known for their ability to treat disease and are supposed to be effective insecticides. The seeds have a high concentration of neem oil that has well known insecticidal and anti-tuberculosis properties and is also used as a lubricant. This incredible plant has been declared the Tree of the 21st century by the United Nations because of these amazing properties and the usefulness it provides to humanity.

Over the years, experts started experimenting on its usefulness in water treatment as it is easily available and economical for rural populations. Some scientists have worked on its antimicrobial property and it appears that the neem oil extracted with ethanol resulted in a substantial microbial reduction of approximately 99% in water from different sources like wells, lakes, rivers *etc.* and some other workers reported a reduction in (Most Probable Number) MPN of tested waters, thereby indicating that it inactivates coliforms. *Salmonella typhi,* the causative microorganism of typhoid fever that is usually prevalent in contaminated waters, was also successfully inhibited by neem oil extract, which clearly points out the significant effect of neem on pathogens.

Although most of the works carried out target *E. coli* or coliforms, specific investigations on other water pathogens will be very advantageous given the fact that neem is used medicinally to treat a host of ailments, as discussed earlier in this section.[31] Therefore, this could be a budding area for future research work. Use of neem leaf powder as a coagulant and for the removal of fluorides from water has been studied albeit to some extent. One of the main disadvantages of neem oil addition to water is that it could make the treated water unpalatable as neem is known to have a bitter taste. Therefore, future work should target the aesthetic appeal of the disinfected water if local communities must drink it. However, this plant does hold promise for developing countries as a natural water disinfectant.

### 2.2.4.5 Okra

Okra or *Hibiscus esculentus* (Ladies' fingers) is yet another useful plant that has amazing coagulating properties that have been widely exploited in the food and pharma industry. The seed proteins and the mucilaginous substance present in them are the main ingredients that are responsible for all its applications. Okra is a plant that is widely grown in many parts of the world such as Asia, Africa and other tropical areas around the globe. Apart from having proteins and mucilaginous compounds, the seeds of Okra are also rich in vitamins and calcium. Okra is commonly used in the household for its nutritive properties and it is usually consumed in steamed, fried or stewed form and is incorporated in soups, salads and curries. Because of its availability in local communities, okra has been looked upon as an option for myriad beneficial applications.

It has been employed to treat some medical conditions such as ulcers and haemorrhoids because of its ability to soothe the body membranes and bring relief. This has been attributed to the mucilaginous substances abundantly present in the seeds and these provide a soothing layer on the surface of the affected parts providing immediate relief. Thus, the medicinal properties are widely known, which ignited some interest to explore its other applications as well. In the food, pharma and cosmetic industries, it has been used as a preservative and lately the mucilage has also found application as a suspending agent and a pharmaceutical adjuvant in drug delivery systems.

In the waste water sector, experts have studied its use as a coagulating agent in the treatment of waste water from various sources such as tannery and other industrial sectors, and they found a remarkable reduction in the turbidity of almost up to 98% in polluted waters, thereby providing an easy and effective solution. In most of the investigations carried out, the coagulating properties of Okra have been highlighted. As safe drinking water should not only have reduced turbidity but should be essentially free of pathogenic microorganisms, researchers have now started focusing on this aspect too.

In a very interesting experiment, a purified Okra protein from Nigeria was tested to ascertain its antimicrobial potency against pathogens in water treatment.[32] It was observed that Total coliforms, *E. coli* and fecal coliforms could be effectively inactivated to the level recommended by the WHO guidelines, which is zero coliforms per 100 ml of tested water. Additionally, the dose required to achieve the inactivation was reduced significantly when purified seed was used. Thus, the purified okra protein was found to be better as compared to the crude okra protein mainly due to the elimination of the organic matter in the seed that was thought to be responsible for shielding the pathogens during water treatment. One of the most important results that was obtained in this study indicates the ability of purified okra protein to act as an effective post treatment alternative. This was mainly because zero regrowth of bacteria occurred even after an extended storage period of treated water (72 hours).

Thus, Okra proteins can be effectively used to prevent bacterial regrowth in distribution pipes like chemical disinfectants that are added to achieve the same target. However, like *Moringa oleifera*, this plant protein also needs to be in a purified form to affect treatment positively by improving disinfection and reducing the disinfection byproducts associated with the presence of organic matter in water. Thus, Okra appears to be a brilliant choice as a coagulant and antimicrobial agent to treat water in developing countries where the plant growth is seen in abundance. More work may be needed to check its antiviral properties, which could be expected soon given the interest that has been triggered globally for economic and safe water solutions for low-income nations.

#### *2.2.4.6 Merits and Demerits of Herbal Approaches*

It is very clear that herbal approaches for water treatment hold great promise for rural communities mainly because they generally use readily available plants and herbs that grow in the local regions and are easy to access. Therefore, there is a reduced dependence on chemicals and other materials that are required to be imported to these regions thereby reducing the transportation costs and hence the overall economics. Most of the studies indicate that the active component of the plant if extracted and used for water treatment provides better results than the crude herb itself. However, this does imply some processing and hence the resultant labour and cost, but a purified herbal active component also minimizes other impurities that would hinder the process.

Thus, future studies in this area are promising. Since there are diverse species of plants that are indigenous to a developing nation, a consistent focus on exploring new species of plants for their specific use in water treatment could be a major way forward. Toxicity related to herbs and plant components when present in drinking water beyond a certain level is a concerning challenge that also needs to be addressed. Therefore, studies along these lines that ascertain the safety of the herbs and their final concentration in treated water will play a pivotal role that will help the drinking water authorities to come up with guidelines for the concentration of the herbal component to be used. Thus, the challenges are many but plant species can surely contribute to advancing the goal of workable drinking water treatment techniques in developing nations.

### 2.2.5 Desalination

Desalination is the process of the removal of salt from sea water and it is one of the oldest methods used by humankind to get pure water from the sea. This has been done since ancient times to increase the quantity of fresh water that is available for various purposes, especially for potable use. It is well established that the concentration of salt varies significantly in fresh water (around 500 ppm) as compared to sea water (approximately

35 000 ppm). Excess salt in drinking water makes it not fit for human consumption due to its unpalatable taste and the resultant sodium fluctuations in the human body post consumption. Therefore, various purification techniques have been practiced on sea and brackish water for a long time. This process that is termed desalination appears to be a promising way of getting pure drinking water to satisfy human needs as sea water is amply available for most of the countries worldwide given the fact that earth is covered with almost 80% water.

The main constraint in desalination processes is the huge amount of energy that is required in the conventional methods using fuels that are adopted routinely. Therefore, these methods are expensive and cannot be readily employed in the developing countries where the economic conditions are poor. Moreover, installing desalination plants is much costlier as compared to water purification techniques that treat ground or surface water. Hence, desalination appears to be a method of choice in regions that are abundant in conventional fuel resources such as the middle east countries. Yet another concerning issue is the emission of greenhouse gases from the use of conventional fuels in the desalination plants. However, despite these demerits associated with the use of desalination, this process appears to be a promising way of addressing the global water shortage mainly due to the vast water bodies in the form of sea across the globe.

Therefore, tremendous efforts have been made by experts and scientists to explore various other renewable energy sources to drive the desalination process so that this technology can be easily and effectively employed in developing countries as well. Before the description of these processes, it is important to understand a little about the conventional desalination techniques so that it will be easier to compare them with the desalination processes that use renewable energy.

### *2.2.5.1 Conventional Desalination Techniques*

The conventional desalination methods that use nonrenewable energy resources are shown in Figure 2.2. The major techniques can be classified into thermal and membrane processes. Thermal processes involve heating up sea water that results in the production of vapour, which condenses to form distilled water that is free of salt. The energy source used in this method could be fossil fuels, nuclear or any other conventional source. Thermal processes can be again classified into:

**2.2.5.1.1 Multi Stage Flash (MSF) Method.** This consists of multiple stages of evaporation and condensation of sea water by flashing a portion of water into steam in multiple stages of counter current heat exchangers; this is the most common method that is employed for desalination across the globe. The main challenge with this type of desalination process is the huge amount of energy required.

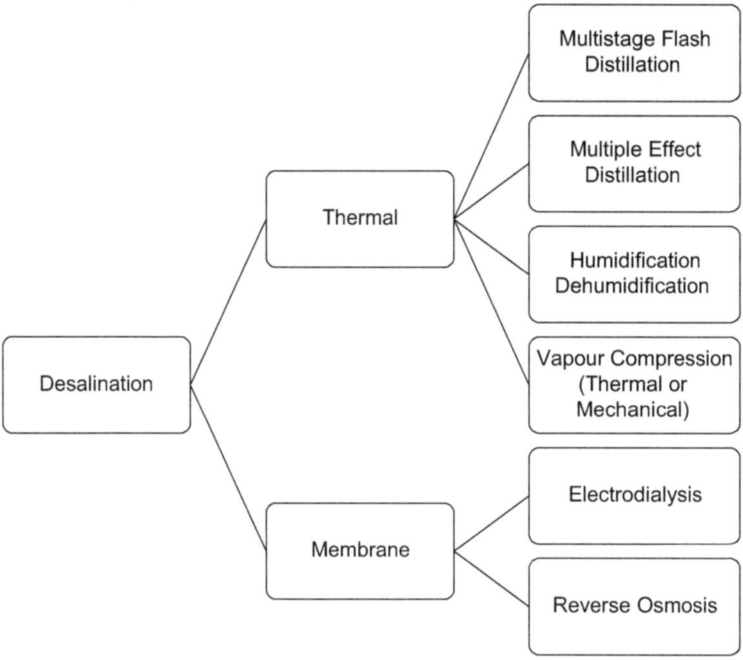


**Figure 2.2** Conventional desalination methods.

**2.2.5.1.2 Multiple Effect Distillation (MED).** This essentially consists of multiple stages or effects and in each stage, the feed (saline) water is heated by steam in tubes. This results in the evaporation of some of the water, producing steam. This steam flows into the tubes of the next stage, again heating and evaporating more saline water. Therefore, each stage essentially uses the energy from the previous stage. Hence, this type of desalination requires considerably less energy as compared to MSF.[33]

**2.2.5.1.3 Vapour Compression (VC).** When the heat obtained from compressed vapour is used in the desalination process, it is called vapor compression and this is done by the latent heat that is available during condensation to generate vapour. This can be brought about by two ways. The first one called thermocompression is carried out by a thermal method using steam and the second method is called mechanical vapour compression, where the vapour is compressed by using a mechanical device such as a compression turbine, increasing the saturation temperature of the steam/vapor if needed.

**2.2.5.1.4 Humidification Dehumidification (HDH).** HDH is much very like the water cycle that occurs in nature. It is a versatile technique and essentially consists of a humidification and a dehumidification zone. The feed water is heated, which results in the formation of vapour, which is mixed with the air in the humidification zone. The humid air is made to contact a cooler surface, which results in the condensation of part of the vapour that is mixed with air in the dehumidification zone. Because of the high energy consumption of these systems, various alterations have been studied that include the use of renewable energy sources to drive the system.

The second major method of desalination is the use of membranes. These processes use semipermeable selective membranes that act as barriers for salts and help in separating the salts from other impurities in a typical desalination process. They may be further classified as follows:

**2.2.5.1.5 Electro Dialysis (ED).** This method is mostly suitable for brackish water distillation and it involves the use of selective ion-exchange membranes. The electrically charged salt ions are separated by these membranes, which results in water that is very low in salinity.

**2.2.5.1.6 Reverse Osmosis (RO).** This is one of the most common and well known methodologies that is adopted in most of the conventional desalination plants. Reverse osmosis can be easily used for both brackish as well as sea water distillation. In this process, the fresh water passes through a semi permeable membrane under high pressure leaving behind the highly concentrated brine solution. It is interesting to note that in recent years, both RO and MSF methods of desalination have gained popularity and their application has augmented by almost 5 to 6 times, respectively, as compared to earlier decades.[34]

As mentioned earlier in this section, the energy requirement of a conventional desalination system is very high and this is the deterrent in its application in developing nations. Experts report that to desalinate sea water with a salinity of 34 000 ppm, the minimum amount of energy needed is around 0.86 kWh $m^{-3}$. However, the desalination plants usually use at least 5 to 26 times more energy than this minimum amount based on the type of processes employed, their maintenance and operation.[35–38] Hence, it is of paramount importance to come up with strategies that could reduce this huge energy consumption and make it feasible in developing countries as well. Very interestingly, global efforts have made renewable energy sources like solar, geothermal, wind *etc.* available so that they can be used to drive the desalination processes. The following section describes various renewable energy based desalination techniques.

### *2.2.5.2 Desalination Using Renewable Energy Sources*

Most common types of renewable energy sources that are used to carry out desalination are described in Figure 2.3.

Some of the main renewable energy sources are wind, solar and geothermal, which have been further divided into sub types. Each major category is briefly described in this section:

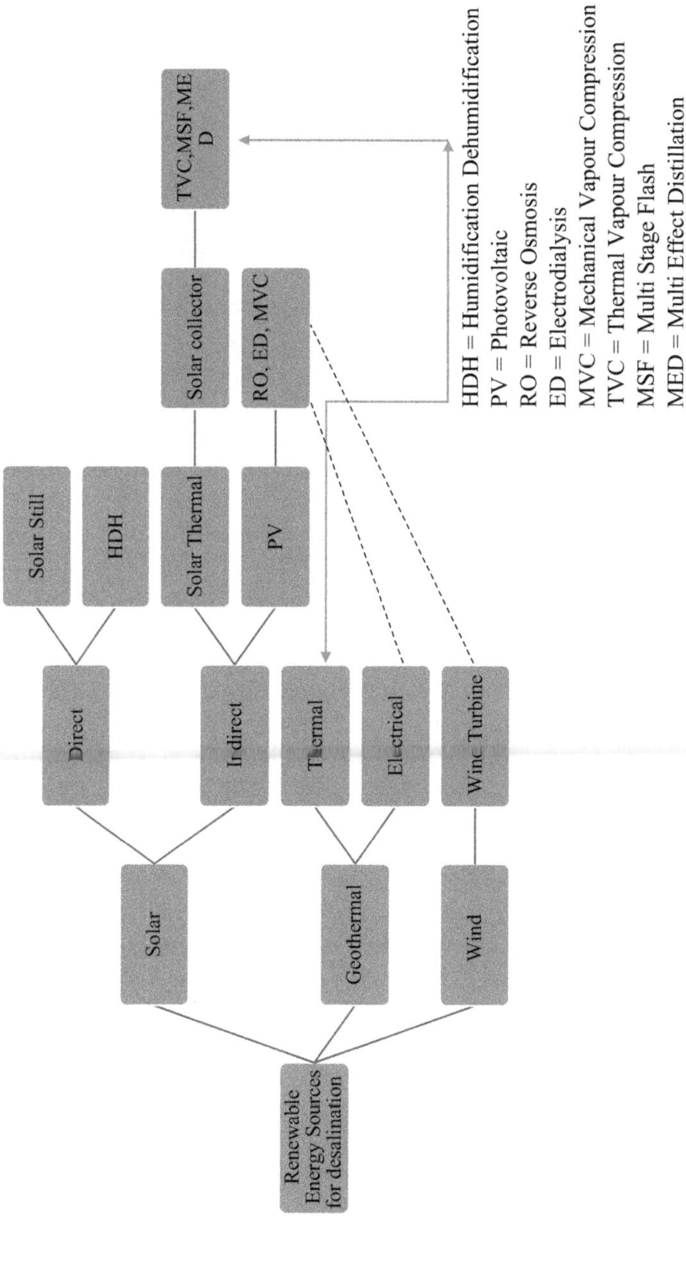


**Figure 2.3** Desalination using renewable energy sources.

**2.2.5.2.1 Solar.** Solar energy is one of the main natural resources that can be harnessed for water treatment and this has been elaborated in the previous sections. Similarly, solar energy has been used to drive desalination processes in several countries worldwide. This can be carried out in two ways, either indirectly or directly (Figure 2.3). In the indirect method, solar energy can be used in two ways. The first method is called solar thermal, where the sun's energy is converted to heat *via* solar collectors, which is then used in the desalination process such as the TVC, MSF or the MED method. The second method is photovoltaic (PV), where the solar energy is converted into electricity using semiconductors that exhibit a photovoltaic effect.

A solar inverter is also used in such systems to change the electric current from DC to AC. This electricity is used to power desalination systems such as RO, ED and MVC. The direct method of using solar energy in desalination involves the use of solar stills, where saline water is kept inside a glass covered chamber, which is pervious to solar rays. Water inside the chamber undergoes evaporation due to solar irradiation and water vapour condenses to form fresh water droplets that get collected separately.

Yet another direct usage of solar energy can be adopted in the HDH system, where the energy for heating is derived from the solar radiation.[39] In recent years, several improvements in solar stills have resulted in enhanced absorption of solar radiation. One such innovation is the use of a nanostructured carbon fibre based material, which floats on the water to be treated. This carbon substrate has excellent broadband light absorbing properties, porosity for capillary action of water, enhanced heat localisation and low thermal conductivity. When solar rays are incident to the substrate, they get absorbed by the substrate and are converted into heat, which is transferred to the surrounding water molecules. The heat causes the water molecules to evaporate and form steam. The resulting steam is condensed to produce pure water leaving waste substances in the solution.

An important aspect in this treatment process is the constant contact (substrate floats above water) of the substrate with water molecules that results in efficient heat transfer. Moreover, the porosity of the substrate helps in the capillary action of water. Simulated incoming sunlight in the intensity range of 0.8–0.85 $KW\,m^{-2}$ using a Fresnel lens resulted in a focused intensity of about 2 $kW\,m^{-2}$ and evaporation efficiency of 2 $Kg\,m^{-2}\,h^{-1}$. A prototype that treats waste water based on this mechanism has been patented that produces about 4–6 L of treated water and the prototype is estimated to cost around $15. It is interesting to note that the waste water can be replaced by sea water in the prototype and treated to obtain pure water.

**2.2.5.2.2 Wind.** Wind is yet another renewable energy source that can be used to power desalination plants. It is a very good option in regions having good potential for wind energy like islands, hilly terrains and regions at higher altitudes. It is essentially a very simple process where seawater is turned to fresh water using windmills. The windmills produce electricity that is stored in batteries and then used to activate the desalination process technology such as reverse osmosis, electrodialysis and mechanical vapour compression. Although the process is simple, the main drawback of this renewable energy source is that wind energy is unpredictable and of intermittent nature so one cannot have the assurance of a constant supply of wind. Moreover, wind supply is dependent to a considerable extent on the topographical location of the area.

Therefore, desalination processes cannot solely rely on wind energy and must have a backup plan. Most experts recommend that wind energy be combined with other renewable energy sources such a solar energy to mitigate this problem. Yet another demerit is the fact that usually, wind energy is converted to electricity and stored in batteries and later converted back to mechanical work. This means that a lot of energy is lost and batteries add to the capital cost. Efforts to mitigate this issue have led to the improvement of the existing process such as the use of windmills to drive the high-pressure pump directly to store fresh water instead of electricity.[40]

**2.2.5.2.3 Geothermal.** The interior of the earth (around 400 miles below earth's surface) is a huge store house of highly pressurised systems consisting of geo pressurized brines, dry rocks, and molten magma, which together can be called geothermal fluid. This geothermal fluid is at an enormously high temperature and when it is used to produce fresh water from saline water, the process is called geothermal desalination. One of the most important aspects of this type of energy is its stable nature; it can produce a consistent supply of energy throughout the day as compared to other renewable energy sources such as solar and wind power. This is primarily because the ground temperature beneath a certain depth remains constant throughout the year and geothermal wells below 100 m can be easily used to drive desalination plants.

Yet another major advantage is that geothermal heat pumps driven by fossil fueled electricity are reported to reduce $CO_2$ emission by at least 50% as compared to fossil fuel driven boilers. It is interesting to note that experts state that 39 countries situated in Africa, south and central America and the Pacific can potentially utilize geothermal energy for their needs. The basic process essentially consists of pumping cold water into a geothermal reservoir that heats it up. The hot water is pumped back and heats up another fluid *via* a heat exchanger, which generates a vapour of that fluid. These vapours run a turbine that generates electricity. This is geothermal energy that can be used to power a desalination plant.[41]

**2.2.5.2.4 Wave.** A simple yet remarkable method of deriving renewable energy is to use ocean wave technologies. This method uses the up and down motion of waves to generate electricity that can be used to power desalination plants. Essentially, the energy in the waves comes from the movement of the ocean caused by the wind and this generates kinetic energy that is incredible. Most importantly, wave energy will never be depleted although it can differ in its intensity. Moreover, it is available everyday year-round, which makes it an attractive choice as compared to wind energy (wind velocities change during the day and night) and solar energy (available only during daytime when there are relatively no clouds). Thus, wave energy production can be considered more consistent as compared to wind and solar energy.

Experts report that wave energy technologies in the coming decades will prove to be as economical as wind driven turbines, which produce energy at approximately 4.5 cents kWh. Thus, wave powered desalination has immense potential for application in developing countries that have high energetic waves available for use.[42] Like other renewable energies, wave energy can be used to either generate electricity, which is then used to drive desalination plants, or the wave powered unit can be directly coupled with a desalination unit. Usually, the wave energy is converted to electricity before use and this can be achieved in three different ways (Figure 2.4).

In the first case, float or buoy systems that are fixed to the ocean floor are used and this device rises and falls with the ocean waves to drive the hydraulic pumps. Basically, this movement causes the movement of a rotor stator assembly in a linear generator and makes electricity that is them brought ashore by underwater power cables. This electricity is used to drive desalination plants. In the second case, an oscillating water column device is employed that is fitted with an air turbine at the top. When the waves rise, the water fills the column and when the wave recedes, the column gets emptied. Due to this process, the air in the column is subjected to compression and gets heated, thus generating energy. This energy is harnessed and sent to the shore *via* an electrical cable and can be used to power a desalination system. Finally, in the third scenario, a tapered channel is used. which is also referred to as a "tapchan" system.

These systems are basically used to capture waves into a raised shore mounted structure, which is generally 3 m above sea level. Water flow out of this reservoir is used to generate electricity, which can be employed to drive a desalination plant. Although there have been several attempts to use wave power to operate desalination units, it is still reported as lagging behind

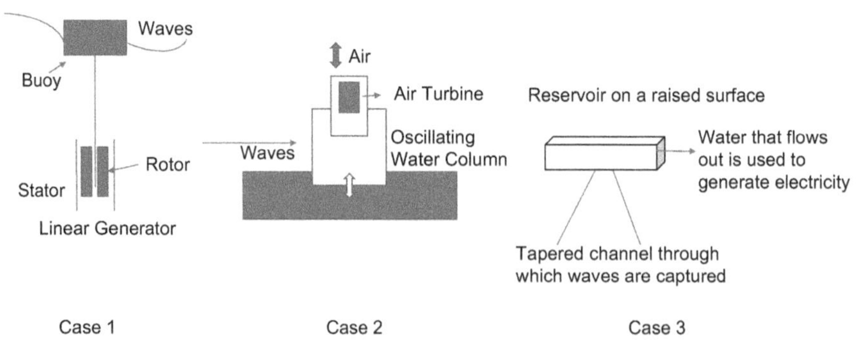


**Figure 2.4** Conversion of wave energy to electricity.

wind and solar energy in terms of commercial development. However, wave energy has far more potential as a resource as compared to wind or solar due to its high predictability. Wave energy has almost 1000 times the kinetic energy of wind and hence much smaller and more compact devices can be used to produce the same amount of power in comparison with a windmill that is huge and occupies space. This reduces land and infrastructure costs and hence is a viable option for developing nations as these devices can be easily integrated with a desalination plant, resulting in power generation at the spot of requirement.[43]

#### *2.2.5.3 Summary on Desalination*

With the growing population and increasing water demands globally, there is always a need to find effective ways of providing safe drinking water and this is a major challenge in developing countries since the technique used should also be economical. Desalination is a very ancient technique that can be easily used to get fresh water in developing nations provided the desalination plants are driven by renewable energy sources. Desalination is practiced globally with the Middle East countries having the maximum amount of desalination plants. Other countries such as Spain, Australia, India, Japan, China and the UAE also use desalination for their fresh water needs. Some examples of developing countries using desalination for obtaining fresh water are listed in Table 2.4.

A huge amount of research in this domain has resulted in the investigation of non-conventional energy sources such as solar, wind, wave and geothermal alternatives that have been described at length in the previous sections. It can be concluded that desalination can be a method of choice when the developing region is located close to the sea like islands and those countries that are close to the coast. Since these projects have major infrastructure and operating costs due to the high energy requirement, developing regions having solar energy, geothermal spots, wave power (being close to the sea) and/or high wind potential can consider using these green and environmentally friendly techniques of powering the desalination plants that are also cost effective. Although there is ample research in these areas, novel technologies that are simple and useful both as small and large-scale methods are required to face the future water needs of developing countries.

### 2.2.6 Emerging Novel Methods

#### *2.2.6.1 Drinkable Book*

Books are our best friends, a fact indeed, but imagine if it could also purify your drinking water! That is exactly what has been done by a remarkable innovation in water purification called 'The Drinkable Book'. The Drinkable book was conceptualized with the main aim to prevent diarrheal diseases that occur by drinking water contaminated with bacteria such as *E. coli.*

**Table 2.4** Some examples of desalination plants that use conventional and renewable energy sources in developing countries.

| S. no. | Desalination method | Developing country | Capacity/cost | Salient features | Reference |
|---|---|---|---|---|---|
| 1. | Multiple effect distillation-thermal vapour compression | Akrimota, Kutch, Thermal Desalination, Gujarat, India | 2×1200 $m^3$ per day | Commissioned in the year 2005 | 29, 33 |
| 2. | Distillation | Republic of Marshall Islands | 100 $m^3$ per day US$3.20 $m^{-3}$ | Commissioned before 1996. Located at Majuro Hospital water supply | 30, 34 |
| 3. | Reverse osmosis | Alanthalai in Thoothukudi District | 60 000 $m^3$ per day | In progress | 31, 35 |
| 4. | Membrane filtration | Delhi Games Village, Water Treatment Plant, India | 4545 $m^3$ per day | Commissioned in the year 2010 | 32, 36 |
| 5. | Reverse osmosis | Panipat Refinery – Effluent Recycling Plant, Panipat, Haryana, India | 21 600 $m^3$ per day | Commissioned in the year 2006 | 33, 37 |
| 6. | Reverse osmosis | Tuvalu, Pacific Island | 65 $m^3$ per day US$89 000 | Commissioned in 2006 at Funafuti to combat drought conditions | 34, 38 |
| 7. | Reverse osmosis | Nauru, Pacific Island | 120 $m^3$ per day >AU$5 $m^3$ per day | Working since 2010. Power station water supply | 30, 34 |
| 8. | Reverse osmosis | Kiribati, Pacific Island | 110 $m^3$ per day | Working, 2002, located at Betio to combat drought | 30, 34 |
| 9. | Reverse osmosis | Tonga, Pacific Island | 16 $m^3$ per day TOP$90 000 | Operational since 2009 at naval base emergency | 30, 34 |

| | | | | | |
|---|---|---|---|---|---|
| 10. | Stepped solar still | Egypt | 6 l per day cost 0.031$ per litre | Modified still using external and internal reflectors could yield 125% more water than the conventional still | 35, 39 |
| 11. | Solar thermal desalination | El Hamarawein, Egypt | 54 $m^3$ per day | 10 kW heat engine used 3500 ppm brackish water treated | 35, 39 |
| 12. | Wind based desalination coupled with RO technology | Greek island | 3360 $m^3$ per day. US$2.02 $m^{-3}$ | Operating since 2007 | 36, 40 |
| 13. | Geothermal desalination | Queensland | 1000–1 00 000 $m^3$ per day US$0.73–1.46 $m^{-3}$ | | 37, 41 |
| 14. | Geothermal desalination & multiple effect distillation | Kimilos and Milos Islands, Greece | 80 $m^3 h^{-1}$ US$1.69 $m^{-3}$ | Pilot project, 470 kW power generator unit. Nearly 7% of the geothermal energy is used to generate electricity from the plant. | 37, 41 |
| 15. | Wave powered desalination based on RO | Vizhinjam, Thiruvananthapuram, in Kerala, India | 10 $m^3$ per day | First wave powered indirect desalination. Since 2004. Meets the water necessity of the fishing community in the locality. | 38, 42 |

This patent pending technology plays a double role by acting as a water filter and an instruction manual. The book (Figure 2.5), around 1 to 2 inches thick, essentially consists of about 20 pages, each of which is about a millimeter thick and is coated with silver nanoparticles. These nanoparticles are deposited by microwave assisted in-situ reduction of silver nitrate on cellulose fibers of an absorbent blotting paper sheet.[44]

The inspiration was drawn from ancient history where our ancestors stored water in silver pots that kept the water clean owing to the antibacterial action of silver. Laboratory tests revealed that the sheets showed antibacterial action toward suspensions of *E. coli* and *E. faecalis* with log reduction values of over log 6 and log 3, respectively.[45] Moreover, the silver leaching from the sheets into water was under 0.1 ppm, which was well within acceptable limits for silver in drinking water as laid down by the US EPA and WHO. Water safety tips are printed in nontoxic food grade ink on each sheet and include information about the need and importance of washing hands and keeping trash away from water sources to prevent contamination. The information is printed in English and local languages as well.

It is interesting to note the simplicity of its use wherein one must just tear a single half sheet of paper from the book and insert it into a plastic box that is provided in the kit. In fact, this plastic box forms the cover of the drinkable book. The water to be treated is passed through this box containing the filter paper and treated water is collected from below. 99.9% bacterial reduction has been reported, which is well within the US EPA guidelines. The book is designed to provide clean water for about a year depending upon its usage and it is very economical as each piece of paper costs only about 10 cents. The entire kit has been tested in the field in South Africa and is being tried in other places such as Ghana and Kenya. An NGO, called 'pAge Drinking Paper', was formed to promote the project further, and it aims at raising funds for scale up and hopes to provide the drinking paper for other projects too.

The Drinkable book is a unique way of providing simple technology with an educational element to educate and purify water. It is promising especially for rural populations as it does not require electricity and can be easily

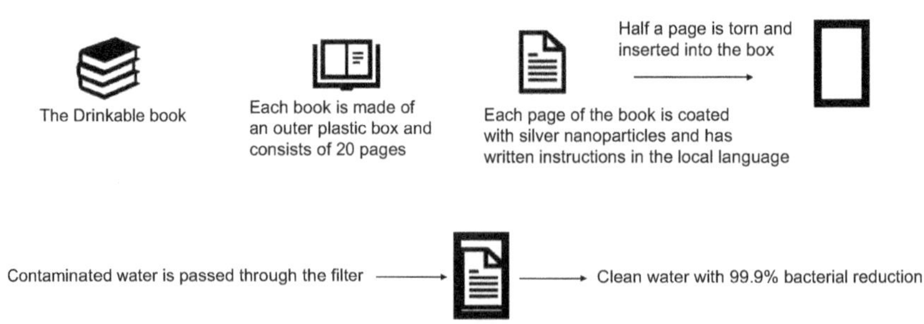


**Figure 2.5** Drinkable book.

used by the locals. Cost of silver, scale up issues, possible clogging of the filter paper, reusability and leaching of silver during long term use could be some deterrents in the technology. Nevertheless, The Drinkable book appears to be a silver lining and provides hope for safe water in times to come.

### 2.2.6.2 *Lifestraw*

One of the most remarkable innovations in water filtration for developing countries is 'Lifestraw' technology, which was launched around the year 2005 primarily as a solution to combat water contamination after natural disasters. It was designed by the Swiss company Vestergaard that manufactures public health tools for people in developing countries. Originally, the filter was developed to remove Guinea worm larvae from contaminated water. Humans are infected by the Guinea worm disease (GWD) or Dracunculiasis when they drink water that contains water fleas that are infected with guinea worm larvae. The filter, initially in the form of a cloth, was later developed into a pipe and more than 37 million lifestraw guinea worm filters used globally have nearly eradicated the disease. Due to the huge success story with guinea worms, a new product was made that could remove almost all the microorganisms present in water.

Thus, in the year 2005, the 'lifestraw', a straw like filter, was designed specifically for personal use for the population in developing nations. In order that the lifestraw technology remains effective for several uses, innovation in the product design was initiated that led to its adaptation in household water filtration to meet individual needs as well as for community use to purify larger volumes of water. These high-volume purifiers called lifestraw community purifiers have been used for filtration in schools and clinics, and other products such as the Lifestraw Go refillable bottle and the lifestraw Mission, which is also a high volume collapsible bag, are very apt for travelers on hikes and expeditions.

Over the years, these products have been used and implemented in projects in over 64 countries globally. It is important to note that these innovations have the potential to augment progress towards the Millennium Developmental Goal for providing access to safe drinking water. Moreover, the Cochrane Collaboration (Cochrane gathers and summarizes the best evidence from research to help make an informed choice about treatment) reviewed 38 randomised controlled trials of various water quality interventions such as chlorination, SODIS, flocculation and disinfection and improved storage of treated water. The basic target was to inactivate microbes and prevent diarrhea. These trials covered more than 53 000 subjects from 19 countries over 20 years and it was reported that among all the interventions studied, filters were consistently the most effective in preventing diarrhea, with an average of 63% reduction. The review pointed to an important finding that it is not enough to treat water at the point of source but it must also be made safe at the point of consumption and the Lifestraw satisfies this need remarkably.[46]

**2.2.6.2.1 What Is Lifestraw and How Does It Work?.** Lifestraw essentially consists of a simple plastic tube that is approximately 9 inches long and 1 inch in diameter. It weighs only about 2 ounces (57 grams) and is easily portable. It can be easily used to filter water and drink directly from an open water body like a river, lake or even a container like a bottle or a glass. The usage of lifestraw is essentially very simple and only requires the person to dip it into water and suck up filtered water directly into the mouth just like one would from a straw. When water enters the lifestraw apparatus (Figure 2.6), the hollow fibers present inside the filter body trap the pathogens as the pores of the membrane are less than 0.2 microns, while clean water passes through and is sucked up by the user. After use, the user must blow air regularly though the device to keep the filter clean and prevent it from clogging.

The lifestraw can filter up to 1000 liters and therefore its lifetime will depend on its usage. For instance, if it is used every day for all the drinking water needs of an individual, it may last for approximately one year but if it used occasionally, it is reported to last longer.[47] Fascinatingly, the device does not filter water when the product has reached the end of its working life and this is indicated very simply by the inability of the user to sip water through the lifestraw. This clearly means that the user is not exposed to the risk of consuming unsafe water inadvertently. It has been reported that the advanced microfiltration technology used by the device removes up to 99.99% of bacteria such as *E. coli, Vibrio cholerae* and *Salmonella typhi* and 99.9% of protozoan cysts like *Cryptosporidium parvum* and *Giardia lamblia*.[48] Thus, Lifestraw greatly reduces the risk of several water borne infections. Moreover, newer product designs such as the Lifestraw mission and the lifestraw family are also able to remove chemicals, salt, heavy metals and viruses as well, thus making them a highly suitable devices for developing countries.

Lifestraw also reduces turbidity by filtering particles of about 0.2 microns and therefore clear water that is free from biological pollutants can be easily obtained. The device is safe since it is free of chemicals and Bisphenol A (BPA is a chemical ingredient commonly used in the manufacture of food and beverage bottles; when leached into water, it is known to cause toxic health effects on the human nervous and hormonal systems) and meets the

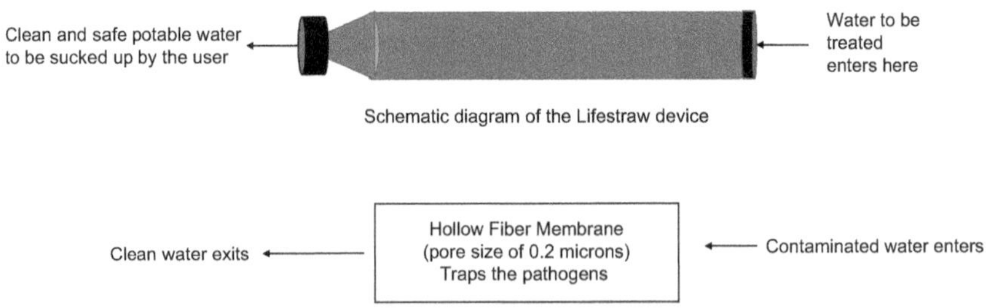


**Figure 2.6** Working of lifestraw.

**Table 2.5** Effect of various lifestraw products on water pollutants when used in developing countries.

| S. no. | Type of lifestraw product | Developing country | Effect on pollutants | Reference |
|---|---|---|---|---|
| 1. | Lifestraw family | Remote rural Congo | Very high antimicrobial effect was seen against Thermo tolerant Coliforms (TTC). 64% of the filtered water was free of TTC and 27% had levels between 1–10 TTC/100 ml | 43, 47 |
| 2. | Lifestraw family | Kenya | High antimicrobial efficacy was reported. 70% or more of water samples filtered using the LifeStraw® Family was free of fecal coliforms. | 45, 49 |
| 3. | Lifestraw family | Intertek Vietnam | Bacterial removal of >99.9999%, virus removal of >99.99%, protozoan parasite removal of >99.9% was obtained.<br>It was found to comply with the USEPA maximum admissible chlorine residual levels of $<4\ mg\,L^{-1}$ and with the maximum product water permissible turbidity <0.5 NTU (NSF/ANSI 63). | 45, 49 |
| 4. | Lifestraw family | Rural Cambodia (Pursat and Kandal Province) | All filters tested removed >99.99% of bacteria (total plate count, total coliforms, thermotolerant coliforms and *E. coli*). High turbidity removal efficacy was also observed. | 45, 49 |
| 5. | Lifestraw family | Ethiopia | High antimicrobial efficacy, extremely high-water aesthetics improvement such as turbidity removal and improvement of taste | 45, 49 |
| 6. | Lifestraw family | Water tested was laboratory spiked water | >Log 6/4/3 reduction in bacteria, virus and parasites, respectively, was obtained. | 44, 48 |
| 7. | LifeStraw | Conducted as lab test in Department of Soil, Water and Environmental Science, University of Arizona, USA (2010) | Log reduction values obtained were:<br>*Escherichia coli* >7.3 (EPA std = 6.0)<br>*Cryptosporidium* oocysts >3.9 (EPA std = 3.0)<br>99.6% turbidity was removed during the test | 45, 49 |

United States Food and Drug Administration (US FDA) and US EPA regulations.

The Lifestraw system requires no electrical power, batteries or replacement parts and therefore it is ideal for the rural population of low income countries.[49] Most importantly, the product has been assessed in several labs and has been subjected to many field tests across myriad developing countries and all the reports have been very promising, as shown in Table 2.5. It has been found to remove almost all bacteria and turbidity, thus improving the taste and color of water.[46] Thus, Lifestraw, a lightweight and slender device with a mighty capability of producing safe potable water for the developing world, is a very useful innovation that is making its presence felt globally.

### 2.2.6.3 *Solar Ball*

A simple yet unique innovation in water purification for developing countries is the solar ball. As the name suggests, this device is spherical in shape (Figure 2.7). This device, developed by a Monash University graduate, works by using solar rays and has the potential to help reduce water pollution, especially in low-income nations, as it is simple, sustainable and economical. The solar ball can produce up to 3 liters of clean potable water per day, which is enough for an individual. The mechanism is very simple and is based on solar evaporation, which is known to eliminate up to 90% of contaminants. Contaminated raw water is placed within the unit and is subjected to evaporation by the solar radiation absorbed by the spherical unit. As evaporation ensues, the pollutants are separated from the water, generating clean condensation on the roof of the device. The condensation is collected and stored and can be used as safe drinking water. This simple yet effective method of water purification is extremely user friendly and the material of the device is durable and has a weather resistant construction.

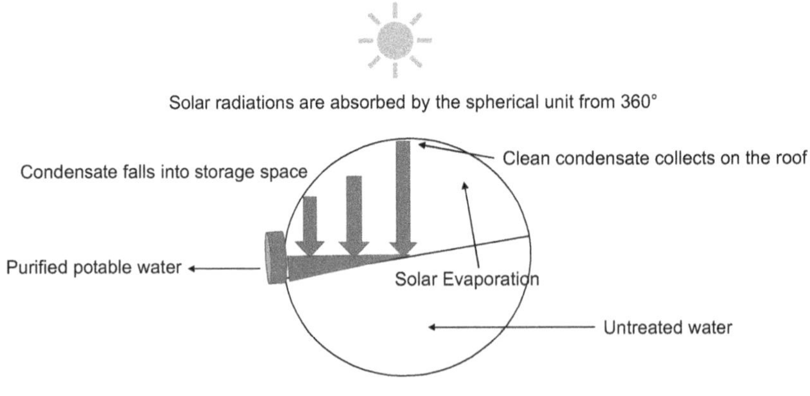


**Figure 2.7** Solar ball.

Therefore, the solar ball is well suited to people from any geographical region with limited access to resources.[50] It is important to note that solar evaporation is an established process and the solar ball is not the only product that uses this technique. However, the solar ball is an effective and unique method comprising a compact unit that can be easily handled. The internal layout and the minute detailing of the product giving importance to specific features such as the opening and locking mechanism, and the slight tilt of the solar ball that ensures that the water flows to the outlet are reported to be well planned.

The spherical shape was designed to capture solar rays from all 360 degrees and to enable it to store solar heat within the device quickly and effectively. Moreover, its unique ball like design makes it easy to be carried and transported. The product also takes into consideration interaction with people from different backgrounds and cultures. The colours and symbols used are neutral and can be easily grasped by locals from any region. An important aspect in the design of any device for water purification is that the material used for making the device should not add to the contamination by way of leaching toxic substances and must be safe and recyclable. It is interesting to note that this aspect of the design has also been taken into consideration and the material used for construction is food and water safety standards complaint and is made of BPA-free polycarbonate. It is sustainable and recyclable and reported to have a long lifespan.[51]

Thus, the solar ball is an innovative initiative by an individual and is one of the many emerging methods to tackle the issue of safe water for developing countries. It is currently in the stage of prototype and lab tests are currently being conducted to improve its possible design and functionalities before it is taken for large scale production. The device will cost approximately \$35 to \$50 per unit and it is hoped to get it to the people of developing nations soon.[52]

### 2.2.6.4 *Water Purifying Bicycle*

Riding bicycles to travel from one place to another is a common occurrence. However, if it could also solve one of the most critical challenges of providing safe drinking water to the people of developing countries who do not have access to clean water, it could really bring about a breakthrough. This was exactly what Japan's Nippon Basic Company had in mind when they introduced the 'Cycleclean', a water purifying bicycle. This exciting innovation was initially designed for use in remote areas and disaster zones during emergencies.

The basic mechanism of the water purifying bicycle is to harness the kinetic energy to purify water and it simply involves cycling to any nearby water body, using it as the source for clean water and producing drinking water by just pedaling the bicycle. The purifying case is attached to the rear seat of the bicycle. Some distinctive features of this innovation are that it can be easily taken to a place that would be otherwise difficult to access by larger vehicles, it does not require an engine or battery for operation and therefore it is suitable for remote places without access to electricity or in emergency situations such as natural disasters.

The water purification system attached to the bicycle is operated by simply pedaling the bicycle and requires no expert knowledge. Additionally, the cylcoclean is powered by riding the bicycle and the clean potable water can be transported to any desired location. The water purification system used comprises four filters, one pre-filter and three filters consisting of a primary filter, activated carbon filter and a MF hollow fiber filter. The water pressure developed by peddling is used to overcome the pressure-drop of these three filters. The prefilter removes the coarse particles and the chemical and physical pollutants are removed by the primary and activated carbon filters. The hollow fiber filter successfully removes the microbial contaminants that may be present in the raw water, thus rendering it free of water pathogens (Figure 2.8).

The water purifying bicycle is reported to produce 6 liters of potable water per minute and produce water for 150 people in an hour. It is currently priced at about US$6600, which is expensive for use in developing countries. To bring down the cost of production of the bicycle, thus enabling it to be used in areas such as Bangladesh in South Asia that has a huge water crisis, the company is putting in efforts to assemble the bicycle in Bangladesh and then fit it with the water purification unit. This is expected to decrease the cost of the bicycle to at least half that of its current price and the delivery charge from Japan will be substantially reduced. Moreover, local production will also provide employment for the people in the region.[53] Thus, given the fact that over a billion people worldwide do not have access to clean drinking water and there are many deaths associated with drinking unclean water, such an innovation is of significant use. Even if the villages have water pipelines and distribution systems in place, it is often difficult to access

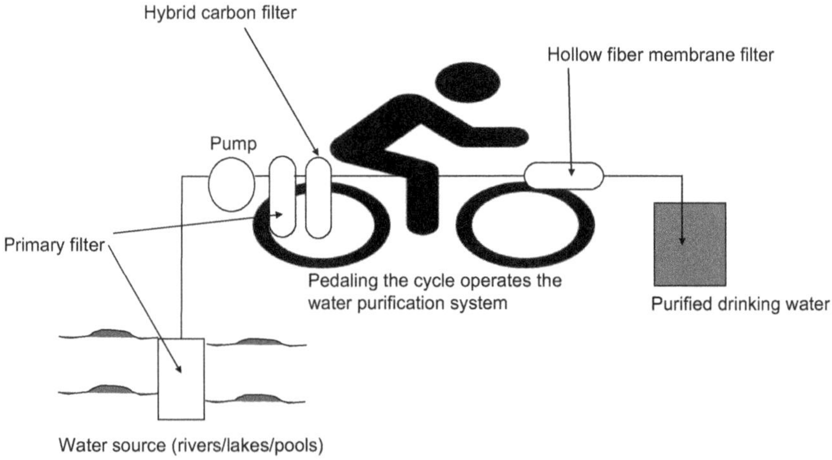


**Figure 2.8** Schematic diagram of water purifying bicycle.

these areas by large trucks. Women must endure walking long distances to procure water for their daily cooking and domestic needs. Challenges such as these can be easily tackled by using these cycloclean devices.

### *2.2.6.5 Life Sack*

This is yet another unique solution for water purification in developing countries. It is well established that many charitable institutions and NGOs reach out to people in remote areas with food grains year-round and most importantly during a natural disaster or crisis in developing nations. It is really fascinating to note that this was viewed as an opportunity by a group of three inventors, to provide pure drinking water as well. In other words, the regular sacks that supply the food grains are replaced by 'Life sacks', which purify water. After the food grains have been used or stored by the locals who receive them, the life sack can be reused to purify water.

Life sack is essentially a plastic bag made of PVC material that is durable for consistent usage (Figure 2.9) and that allows solar radiation to penetrate the contaminated water present in it. Therefore, the UV-A radiation and the bag's thermal treatment process work together to inactivate most micro-organisms present in water. It is a well-known fact that UVA radiation and thermal treatment are individually harmful to the microorganisms commonly found in water. Hence, it may be expected that the combined effect will be detrimental to the water pathogens and increase the efficiency of the process.

The life sack also contains a filtration system of 15 nm embedded into it, which is capable of removing very small size bacteria and delivering clean and safe water. Moreover, the bag can be easily worn as a backpack and allows easy movement to and from the source to the point of use. It can carry up to 20 liters of water, which is a significant amount and can easily cater to the needs of a family.[54] Thus, by merely replacing the conventional bags, it

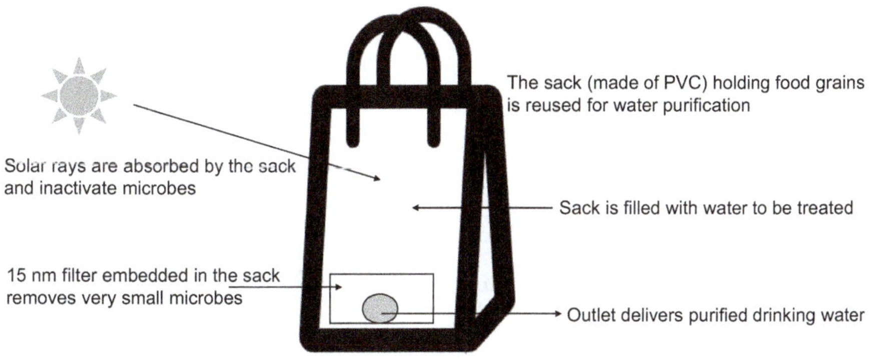


**Figure 2.9** A schematic representation of the life sack.

was not only possible to conveniently supply food to remote regions in developing countries, but also bring about an amazing yet simple method for water purification in low-income nations.

### *2.2.6.6 Atmospheric Water Generators (AWGs)*

When it comes to safe drinking water for human needs, there are two situations that are possible in the global scenario. People may have availability of water even if it is contaminated so that there is a possibility of purifying it before consumption. However, there are people who do not have any access to water (polluted or clean) without travelling long distances. It is a well-known fact that women from rural villages must travel very far to fetch water to their homes and such collected water usually requires some form of treatment before it can be safely consumed. Drinking polluted or insufficiently treated water is detrimental to human health and is also responsible for the huge number of diarrheal deaths worldwide. Atmospheric water generators are very apt for such situations as these devices extract the humidity present in the air itself and deliver purified water to the end user. This eliminates the need for any water source such as municipal supply, lakes, or wells and therefore the necessity to travel to fetch water.

Essentially, in an AWG, after extracting the humidity from the air, the water vapour is condensed in a cooling chamber and the collected water is rendered potable by passing it through a series of filters and/or an ultraviolet chamber (Figure 2.10). One major drawback of this process is the huge energy involved but there has been some research to develop AWGs that produce potable water at reduced energy cost. One example of an AWG product is the device launched by the company Ecoloblue during the time of the California drought in the year 2014 and it proved to be a valuable tool to procure safe drinking water when other sources of water were unavailable. This device has many customers in the US and is also distributed worldwide.[54]

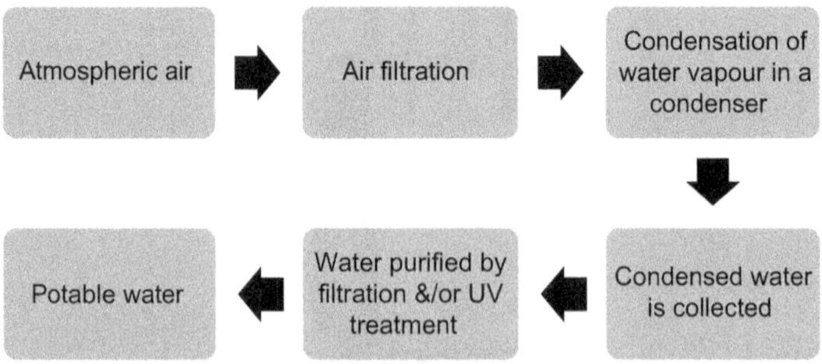


**Figure 2.10** Atmospheric water generators.

### *2.2.6.7 Pure Water Bottle*

A water bottle is the most common thing that any traveler packs during long journeys and hikes, especially to remote places where there are scarce water resources or even if water is available, it may not be fit for consumption. A very common technique to purify water on the go is to use chemical disinfectants such as chlorine and iodine tablets as these are known to be antimicrobial in nature and they are easily available. However, the treated water is not palatable and the process also takes at least 30 mins. Thus, alternate solutions have been explored by several innovators and one such novel method is to use the 'pure water bottle'. This bottle filters contaminated water in 2 minutes and uses a combination of micron size water filters along with a UV light system (Figure 2.11).

Essentially, the bottle comprises an inner and outer chamber that can be separated. The filter and the UV system are attached inside the inner chamber. The top outer part of the inner chamber contains a central knob and an outlet. To operate the bottle, untreated water from any source such as lakes, rivers *etc.* is filled in the outer chamber. The inner chamber is plunged into the outer chamber that results in filtration of the untreated water through the 4-micron size filter. When the central knob on the top is turned by hand, the UV system that is inbuilt is activated and pure potable water is made available through the outlet. This synergism of filtration and UV exposure removes almost 99.9% of pollutants from any water source. This is a very good device that can be used in developing countries.[54] This is a prototype that has been well received and it is expected to be manufactured in large numbers soon.

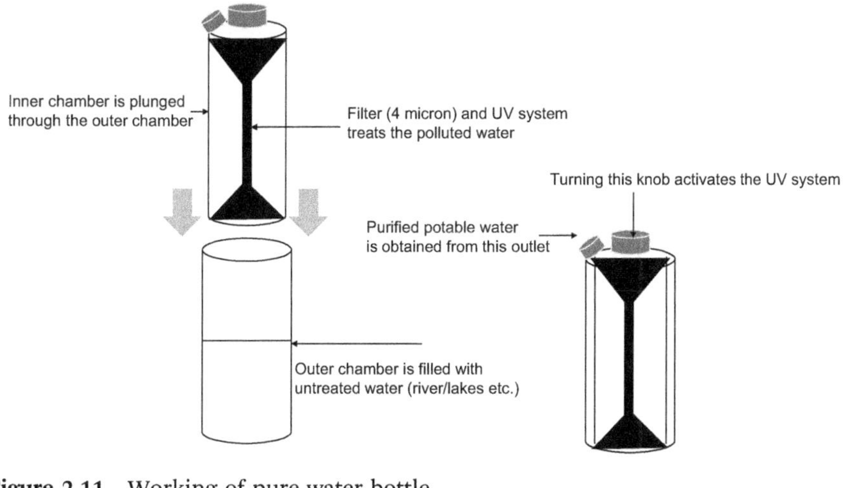


**Figure 2.11** Working of pure water bottle.

### 2.2.6.8 Hydrodynamic Cavitation

Cavitation is a fascinating phenomenon seen in nature. Snapping shrimps that commonly inhabit coral reefs use this phenomenon for prey hunting and communication. The shrimps use a special shaped claw to produce a cavitating high speed water jet that can paralyze their prey. While these shrimps have used cavitation for a meaningful need, the detrimental effects of this phenomenon can also be noticed in nature and one such effect is the erosion of propeller blades of ships. Thus, any phenomenon in nature can have multiple uses, some that could be undesirable and others that can be utilized for constructive purposes and cavitation is not an exception. Interestingly, some research groups across the globe have found ways to harness the positive effects of hydrodynamic cavitation.

Cavitation can be described as a remarkable process that results in the nucleation, growth and collapse of gas filled cavities in a liquid medium. This can be attained by alteration in the flow and pressure of the liquid as it passes through a narrow constriction and is called 'hydrodynamic cavitation' or by the passage of ultrasound in which case it is called 'acoustic cavitation'. The focus here is mainly on the hydrodynamic cavitation, which is considered an advanced oxidation process (AOP). The use of hydrodynamic cavitation is diverse, and it has been effectively applied in the field of waste water treatment, extraction, crystallization and water disinfection. The history, principle and mechanisms of hydrodynamic cavitation and its role in microbial inactivation have been extensively elaborated in case studies 5 and 6 that appear in chapter 5. Its role as an energy efficient and cost effective water disinfection process has also been described in these case studies.

It is indeed fascinating to note that hydrodynamic cavitation can not only be used in special reactors that have been designed for lake water treatment, as seen in case study 5, but what is even more enchanting is the fact that a simple modification in hand pumps that are commonly used to draw water from bore wells in developing nations also results in cavitating conditions. This has been usefully exploited in case study 6 for bore well water treatment and the complete story can be found in chapter 5.

The cost effectiveness of hydrodynamic cavitation has been researched by several experts including the authors. It has been reported that the cost of water treatment by hydrodynamic cavitation using an orifice plate set-up was around US\$1.4 $m^{-3}$ and this was substantially cheaper compared to other treatment techniques such as sonochemical reactors costing US\$14.88 $m^{-3}$ or a high-pressure homogenizer costing US\$6.55 $m^{-3}$. An attractive aspect of using hydrodynamic cavitation is its ability to reduce and/or remove most physical, chemical and biological pollutants and it does not leave behind any disinfection byproducts. Extensive study on hydrodynamic cavitation is underway by experts worldwide including the author to improve its scalability as currently, it can be used only up to pilot scale level. However, with continuing efforts, hydrodynamic cavitation will surely find its place as one of the energy efficient and economical techniques for water treatment in developing countries.

### 2.2.6.9 Pros and Cons of Emerging Techniques

Necessity is the mother of invention; an accepted proverb that appears to be apt for water treatment techniques as well. Innovation is very clearly evident in the remarkable creations that have been described in the preceding section. Although some novel methods might have been inadvertently missed out in the discussion, most of the methodologies covered here are brilliant technologies in themselves that are specifically targeted for developing countries. Most of them have the common features of being extremely simple, efficient and cost-effective techniques that are feasible in rural scenarios. Most of them, except the LifeStraw, which appears to be readily available in large numbers, are still in the prototype stage or are just beginning the journey towards production at mass volumes.

Therefore, factors such as initiatives from NGOs and other institutions, supply to the affected arears, training the rural population and compliance from the end user are critical in determining the success of such emerging techniques. Nevertheless, with an increasing global water crisis, such methods will be well received, and this will also encourage the development of more such simple yet effective innovations for drinking water in developing countries.

## 2.3 Conclusions

This chapter essentially discussed all the major techniques of water treatment that can be easily adopted by developing nations. SODIS, filtration, herbal approaches, desalination and all the emerging new technologies described are most suited for developing countries mainly because most of these methods are simple, efficient and inexpensive. Each methodology has its pros and cons, which are discussed at the end of each topic. Overall, it can be concluded that there is no one universal solution for treating water that will be suitable for low-income regions. However, depending on the region, needs and the level of contamination in the source water, careful selection can be made to provide safe drinking water.

It is very important to note that this chapter dealt with ways of treating water sources that are either groundwater or surface water. However, owing to the extensive water crisis being faced currently, there is a need for additional fresh water resource or ways to add to the existing water bodies. One such method is to conserve water and that can be done by rain water harvesting and treating the harvested rain water to make it potable. Yet another way to combat the water scarcity is to reuse waste water (which can be either direct or indirect potable reuse) after appropriate treatment so that it becomes fit for human consumption. More information about rain water harvesting and waste water reuse is included in the next chapter.

## 2.4 Further Research and Recommendations

- Some new emerging techniques such as the solar ball are in the nascent stage and require scaling up so that they can be made available to a larger population in rural communities.
- SODIS is the most commonly used technology in developing countries and therefore efforts to improve its effectiveness by investigating new additives that can be combined with SODIS will be useful. Further studies in scaling up to provide for larger communities are the need of the hour.
- Filtration using simpler and more easily available natural materials should be investigated to add to the existing knowledge. Innovative techniques for back flushing, prevention of clogging and reusability without compromising the effectiveness of the filter should be investigated.
- Further emphasis on the use of renewable energy sources to drive desalination plants is required and ways to effectively use solar, wind, wave and geothermal power coupled directly to the desalination system without storing it as electricity should be explored to minimize cost further.
- Toxicology studies on the herbs commonly used to treat water should be conducted to ascertain their safely limits and newer medicinal plants should be explored for use in water treatment.

## References

1. https://www.cdc.gov/safewater/solardisinfection.html (last accessed May 2017).
2. K. K. Jyoti and A. B. Pandit, *Drinking Water Disinfection Techniques*, New York, CRC Press, 2013.
3. http://www.sodis.ch/projekte/index_EN (last accessed May 2017).
4. F. Mendez-Hermida, E. Ares-Mazas, K. G. McGuigan, M. Boyle, C. Sichel and P. Fernández-Ibañez, *J. Photochem. Photobiol., B*, 2007, **88**, 105–111.
5. K. G. McGuigan, R. M. Conroy, H. J. Mosler, M. du Preez, E. Ubomba-Jaswa and P. Fernandez-Ibañez, *J. Hazard. Mater.*, 2012, **235–236**, 29–46.
6. M. du Preez, R. M. Conroy and K. G. McGuigan, *Environ. Sci. Technol.*, 2012, **46**, 3036–3037.
7. A. Rose, S. Roy, V. Abraham, G. Holmgren, K. George, V. Balraj, S. Abraham, J. Muliyil, A. Joseph and G. Kang, *Arch. Dis. Child.*, 2006, **91**, 139–141.

8. S. K. Mani, Development and evaluation of small-scale systems for solar disinfection of contaminated drinking water in India, School of Applied Sciences, PhD thesis, Northumbria University, Newcastle-upon-Tyne, UK, 2006, pp. 727.
9. M. B. Keogh, M. Castro-Alférez, M. I. Polo-López, I. F. Calderero, Y. A. Al-Eryani, C. Joseph-Titus, B. Sawant, R. Dhodapkar, C. Mathur, K. G. McGuigan and P. Fernández-Ibáñez, *Sol. Energy*, 2015, **116**, 1.
10. J. Acher and J. Juven, *Appl. Environ. Microbiol.*, 1977, **33**, 1019–1022.
11. A. Cooper and G. Yogi, *J. Sol. Energy Eng.*, 2002, **124**, 305–310.
12. G. Towers, *Nature*, 1985, **131**, 839–840.
13. T. S. Makuba, K. M. Taba, F. Rosillon and B. Wathelet, *Acc. Rendered Chem.*, 2016, **19**, 827–831.
14. http://www.sswm.info/category/implementation-tools/water-purification/hardware/semi-centralised-drinking-water-treatmen-2 (last accessed June 2017).
15. (a) H. van der Laan, D. vanHalem, P. W. M. H. Smeets, A. I. A. Soppe and J. Kroesbergen, *Water Res.*, 2014, **51**, 47–54; (b) M. J. Hedegaard and H. Albrechtsen, *Water Res.*, 2014, **48**, 71–81.
16. T. Asami, H. Katayama, J. R. Torrey, C. Visvanathan and H. Furumai, *Water Res.*, 2016, **101**, 84–94.
17. A. Bhatnagar, W. Hogland, M. Marques and M. Sillanpaa, *Chem. Eng. J.*, 2013, **219**, 499–511.
18. M. L. Sens, M. L. Emmendoerfer and L. C. Muller, *Desalin. Water Treat.*, 2015, **53**, 15–26.
19. M. S. H. Boutilier, J. Lee, V. Chambers, V. Venkatesh and R. Karnik, *PLoS One*, 2014, **9**(2), e89934.
20. A. B. Pandit and K. K. Jyoti, *Annu. Rev. Chem. Biomol. Eng.*, 2015, **6**, 217–246.
21. S. L. Molloy, R. Ives, A. Hoyt, R. Taylor and J. B. Rose, *J. Appl. Microbiol.*, 2008, **104**, 998–1007.
22. H. K. Oh, S. Takizawa, S. Ohgaki, H. Katayama, K. Oguma and M. J. Yu, *Desalination*, 2006, **202**, 191–198.
23. X. Fan, Y. Tao, L. Wang, X. Zhang, Y. Lei, Z. Wang and H. Noguchi, *Desalination*, 2014, **335**, 47–54.
24. P. D. Boateng, 2001, *Comparative studies of the use of alum and Moringa oleifera in surface water treatment*. MSc Thesis, Dep. Civil Eng., Kwame Nkrumah Univ. Sci. Technol., Kumasi, Ghana.
25. S. A. A. Jahn, *Entwickl. Ländl. Raum*, 1989, **23**(4), 22–25.
26. A. Ndabigengesere and K. S. Narasiah, *Water Res.*, 1998, **32**(3), 781–791.
27. A. Olsen, *Water Res.*, 1987, **21**(5), 517–522.
28. T. Bhattacharjee, M. R. Gidde and N. K. Bipinraj, *Int. J. Eng. Res. Dev.*, 2013, **5**(10), 7–10.
29. P. Sundaramurthi, S. Dhandapani, S. Ponnusamy and M. Subbaiyan, *Hitek J. Biol. Sci. Bioeng.*, 2012, **1**(1), 1–7.
30. S. Sa'id, K. Mohammed, D. B. Adie and C. A. Okuofu, *Bayero J. Pure Appl. Sci.*, 2016, **9**(1), 236–240.

31. A. N. Jones, *Arid Zone J. Eng. Technol. Environ.*, 2017, **13**, 6–14.
32. J. Joshi and O. Sahu, *Int. J. Clin. Nutr.*, 2014, **2**, 36–40.
33. Gujarat Reliance Project – IDE Technologies. http://www.ide-tech.com/blog/b_case_study/gujarat-reliance-project/ (last accessed July 2017).
34. S. Manju and N. Sagar, *Renewable Sustainable Energy Rev.*, 2017, **73**, 594–609.
35. Government of Tamil Nadu. Government orders Tamil Nadu Government Portal. http://www.tn.gov.in/go_view/dept/21(last accessed July 2017).
36. Projects. http://www.wabag.com/projects/ (last accessed July 2017).
37. Nuclear Desalination – World Nuclear Association. http://www.world-nuclear.org/information-library/non-power-nuclear-applications/industry/nucleardesalination.aspx (last accessed July 2017).
38. https://gsd.spc.int/sopac/docs/SOPAC%20Technical%20Report%20437%20Desalination%20for%20Pacific%20Island%20Countries.pdf (last accessed June 2017).
39. Z. Omara, A. Kabeel and M. Younes, *Energy Convers. Manage.*, 2014, **78**, 876–881.
40. E. Kondili and J. K. Kaldellis, Wind energy based desalination processes and plants, in Proceedings of the Xth world renewable energy congress; n.d. p. 1–7.
41. M. Goosen, H. Mahmoudi and N. Ghaffour, *Energies*, 2010, **3**, 1423–1442.
42. P. A. Davies, *Desalination*, 2005, **186**, 97–109.
43. http://www.oceanenergycouncil.com/ocean-energy/wave-energy/ (last accessed August 2017).
44. T. A. Dankovich and D. G. Gray, *Environ. Sci. Technol.*, 2011, **45**(5), 1992–1998.
45. T. A. Dankovich, *Environ. Sci: Nano.*, 2014, **1**, 367–378.
46. http://www.lifestraw.com (last accessed July 2017).
47. S. Boisson, M. Kiyombo, L. Sthreshley, S. Tumba, J. Makambo and T. Clasen, *PLoS One*, 2010, **5**(9), e12613.
48. T. Clasen, J. Naranjo, D. Frauchiger and C. Gerba, *Am. J. Trop. Med. Hyg.*, 2009, **80**, 819–823.
49. http://www.wiley.com/legacy/wileychi/selendy/supp/e-brochure-lifestraw-english.pdf (last accessed July 2017).
50. http://www.monash.edu/news/articles/1679 (last accessed July 2017).
51. http://www.abc.net.au/tv/newinventors/txt/s3158420.htm (last accessed July 2017).
52. https://livingnow.com.au/solarball/ (last accessed July 2017).
53. http://www.nipponbasic.ecnet.jp/e-cc.html (last accessed August 2017).
54. http://causetech.net/innovation-zone/tech-inspirations/life-sack (last accessed August 2017).

CHAPTER 3

# *Rain Water Harvesting in Developing Countries*

Chapter 2 explored the central theme of this book, *i.e.* drinking water treatment for developing countries. Various methodologies, their working principles, pros and con and their suitability to rural areas were discussed in detail in the previous chapter, which brought out the need to have a wise choice of water treatment methods based on the location, need of the people and the pollutants in question. This was essentially to treat the ground or surface water already existing in the developing countries. However, from all the preceding discussions in the previous chapters, the looming issue of water scarcity cannot be emphasised more. Therefore, yet another important aspect of water treatment in the developing nations would entail identifying additional new water resources that can be used as potable water sources after appropriate treatment.

Rain water harvesting is one such technique that can be used to preserve water and use it for drinking purposes after adequate treatment to ensure that the quality is good and fit for human consumption. This chapter describes RWH in detail, covering its history and the basic design of the RWH system with elaborate details on the various parts along with the calculations that must be carried out to arrive at the harvested rain water supply. For ensuring that the harvested rain water is free from all types of possible pollutants, it is imperative that the harvested rain water must be suitably treated. Thus, treatment methods such as use of chemicals, UV, SODIS and filtration that are apt for developing countries are detailed in this chapter.

## 3.1 Introduction

Adequate, safe drinking water supply is indispensable to life. However, millions of people throughout the world still do not have access to this necessity. The global population is augmenting rapidly and the issue of

Drinking Water Treatment for Developing Countries: Physical, Chemical and Biological Pollutants
By Aniruddha Bhalchandra Pandit and Jyoti Kishen Kumar

Published by the Royal Society of Chemistry, www.rsc.org

supplying safe and adequate drinking water to meet societal needs is one of the most important and urgent challenges faced by the decision makers. To meet the water demands of a society, generally, there are two approaches. The first approach is finding alternate or additional water resources using conventional approaches and the second approach is utilizing the limited amount of water resources available in a more efficient way. Among the various alternative methods to supplement freshwater resources, rain water harvesting (RWH) is a very promising choice. This can be considered as a simple but effective and environmentally sound measure.

RWH in its broadest sense is a method that is used for collecting and storing rain water for human use. It has been practiced for centuries owing to the spatial variability of rainfall and is an important water source in many areas that have heavy seasonal rainfall. Even in regions where there is a lack of good quality freshwater or groundwater, RWH is an attractive option. In industrialized countries, RWH systems have been developed to reduce water bills or to meet the needs of remote communities or individual households. Rain water harvesting for potable water use is now commonly employed in most countries at different levels for thousands of families. The purpose and basis of the utilization of rain water vary in each country. RWH is very much encouraged in Denmark and the Netherlands and Germany as well. It is interesting to note that the rainfall over the entire territory of Africa can meet the needs of 9 billion people.

However, RWH for water supply has not been widely used due to several reasons such as economic factors. In Japan, the utilization of rain water is widely promoted by the active participation of citizens to prevent flooding and for restoring the groundwater. Singapore is yet another example where RWH is practiced. Currently, 80% of the water used by Changi International Airport is from storm water. The rain water is taken from the airport runway and used for sanitation. In Indonesia, the infiltration of rain water into the ground is carried out to conserve the city's groundwater sources.[1] Basically, RWH can be used for domestic purposes as well as for agriculture and aquifer replenishment. This chapter focusses on RWH for drinking purposes and describes the basic technology with the main emphasis on the treatment of harvested rain water.

## 3.2 Definition

Rain water harvesting (RWH) has been one of the outstanding water conservation methods available to humankind for eons now. This section of the chapter essentially deals with everything about RWH. Although it is established that RWH means the collection and storage of rain water for domestic or agricultural uses, it is interesting to note some of the RWH definitions that are commonly cited in the literature.

1. Rain water harvesting is defined as the collection of water from surfaces on which rain falls, and the subsequent storage of this water for later use.[2]

2. RWH is the practise of capturing, infiltrating or utilizing rainfall from roofs, constructed catchment surfaces, and driveways, sidewalks, parking lots and streets.[3]
3. Rain water harvesting is the collective term for a wide variety of interventions to use rainfall through collection and storage, either in soil or in man-made dams, tanks or containers bridging dry spells and droughts.[4]
4. Rain water harvesting, in its broadest sense, is a technology used for collecting and storing rain water for human use from rooftops, land surfaces or rock catchments using simple techniques such as jars and pots as well as engineered techniques.[5]
5. Access to safe drinking water is currently measured by whether a person has access to an "improved" drinking water source. Rain water harvesting systems are such a type of "improved" water source that over 60 million people across the world primarily rely on for collecting drinking water, and it is likely that hundreds of millions more utilise rain water as a supplementary source of water for both potable and non-potable uses.[6]

## 3.3 History

Water, although a natural resource, has always been a basic need for humankind over the centuries. This can be attributed to various activities such as industrial, agricultural, domestic and augmented pollution that have always made water scarce and contaminated. To solve this crisis, RWH has been practised for centuries so that basic water needs are met for an individual or community. RWH has been employed in industrial and developing countries alike, especially in arid regions. Apart from providing water in arid regions, it also helps in reducing water bills.[7] Historical evidence of RWH is spread worldwide and Table 3.1 lists some of it.

The technology of RWH is deeply rooted in India and an array of different methods are still used today like the talibs (reservoirs that provide irrigation for plants and drinking), Jhalaras (specially constructed tanks used for local communities), Baoris (wells dug into the ground that are used for drinking) and Johads (dams that are used to capture rain water). It is interesting to note that many ancient cities in the Indus Valley civilization used rock vats to collect rain water for the purpose of drinking and agriculture.

Yet another technique used for centuries in India was to build RWH systems on rooftops and this simple technology has spread round the globe today. One of the most impressive RWH systems can be found in Istanbul in the Sunken Palace. During the time of the Roman Empire, several RWH systems with state of the art technology emerged, which evolved right up to the 6th century AD and the rule of Emperor Caesar.

Although RWH systems became popular for providing water, further growth was stalled due to the advent of better techniques such as water pipes and reservoirs in constructed buildings. Additionally, the prospect of

**Table 3.1** History of rain water harvesting.

| S. no. | Region | Time since RWH practised | Mode/application of RWH | Reference |
|---|---|---|---|---|
| 1. | Caribbean & Middle East | Over 300 years | Rooftop catchments & cistern storage | 8 |
| 2. | Honduras, Brazil, and Paraguay | Data not available | Rooftop catchments | 9 |
| 3. | Thailand | Since 1979 | Rain water collection from roofs or gutters into jars | 23 |
| 4. | Asia | Since 10th century | Rooftop catchments & cistern storage | 24 |
| 5. | Farming communities of Baluchistan and Kutch, India | Around 3rd century BC | RWH for agriculture | 25 |
| 6. | In ancient Tamil Nadu | During the later Chola period, (1070 to 1279 AD) | Rain water from the Brihadeeswarar temple (located in Balaganpathy Nagar, Thanjavur, India) was collected in the Shivaganga tank. | 25 |
| 7. | The Vīrānam tank was built in Cuddalore district of Tamil Nadu | (1011 to 1037 CE) | RWH to store water for drinking and irrigation purposes | 25 |

diseases that could spread from the collected rain water also led to the diminished use of RWH systems. However, over the years, due to climate change and global warming, people have started to think about water conservation again. Accelerated interest in RWH has led to the formation of RWH associations globally and the expansion of research in RWH worldwide.[8] This technology has been adapted in arid, semi-arid, rural and urban areas and can serve as a primary or supplementary water source for rising population demands. Individuals and groups all over the globe have developed many variants of RWH systems for use.[9] Moreover, water policies of developing countries list RWH as a source of domestic water supply.

Thus, it is accepted that rain water, a precious gift from the sky, cannot be wasted anymore and can be effectively used for various domestic purposes and also to reduce the financial burden of increased water charges.[10]

## 3.4 Basic Design of RWH Systems

A typical RWH system has the following components.[11]

(i) *Catchment Area*

This is the surface upon which the rain falls. Commonly, it could be a roof that is appropriately sloped towards the direction of storage and discharge. It is usually made of a suitable material such as concrete or asbestos.

(ii) *Gutters*

These are conveyance systems that transport collected rain water from the catchment to the storage tank. These must be designed based on the site, rainfall and roof characteristics.

(iii) *First Flush Device*

An important part of the conveyance system, the first flush device improves the quality of the harvested rain water by diverting the initial amount of rain water away from the storage tank as it is usually dirty and contaminated.

(iv) *Storage Tanks*

These are cisterns or tanks where the collected rain water is safely stored for further distribution as and when needed.

(v) *Conveying*

Essentially, this is the delivery system for the treated rain water either by gravity or pump. This can be simply a container to extract the water from the storage tank or a pipe functioning exclusively as an outlet. At household level, this will be the piping that supplies the building and at the community level, this could be either a single outlet pipe or a network system catering to the needs of multiple users.

(vi) *Water Treatment*

This involves the use of filters and screens to remove solids and particulate matter, and the use of additives to settle, filter and disinfect the collected rain water. Thus, these elements maintain and enhance the water quality.

Figure 3.1 shows a typical RWH system and its basic components.

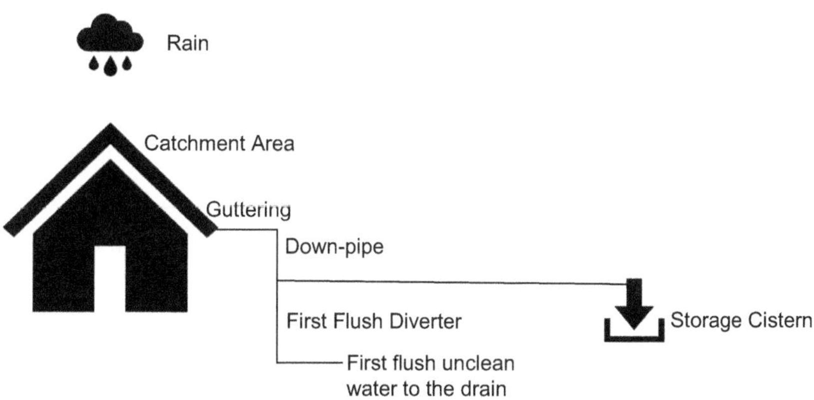


**Figure 3.1** Rain water harvesting system.

## 3.5 A Little More on Each Component

### 3.5.1 Catchment

The catchment area of a rain water harvesting system is the surface that receives rainfall directly and contributes the water to the system. For household water supplies, the roof of the building is generally used as the catchment area. Many other surfaces can also be used such as courtyards, threshing areas, paved walking areas and large rock surfaces as well.[11] Rain water harvested from these surfaces should be used for non-potable purposes due to the increased risk of contamination as compared to the rooftop harvested water, which can be safely used for drinking (rain water harvesting systems). The catchment area and the type of material employed in its construction directly influence the collection efficiency and water quality.

Commonly used materials are corrugated aluminium, galvanized iron, concrete, tiles, slates *etc.* The material used must be non-toxic and must not affect the water quality (rain water guide). Roof painted with lead based paint or an asbestos roof is generally regarded as unsuitable for RWH. However, there are reports that state that the danger of developing cancer from ingested asbestos is very slight whereas the inhalation of asbestos dust is dangerous. Generally, a well thatched roof is usually accepted as safe for RWH. It is important to note that any catchment area must be maintained properly to ensure that the resultant rain water runoff is not contaminated. Sometimes, grease, oil, paints *etc.* are often left on rooftops and these can result in contamination of rain water runoff.

Moreover, due precautions must be taken to ensure that debris from the roof does not enter the tank and the catchment area should be kept clean always, especially during monsoon. Periodic removal of dust, leaves and bird droppings minimizes bacterial contamination and maintains the quality of collected water. It is also advisable to boil the water before consumption for added safety.[11]An important and interesting aspect is the calculation of the catchment area as this determines the RWH water that will be available for use. The 'Rational Method' is generally applied to calculate the amount of rain that can be collected off a roof surface per year. It is given as follows:

$$\text{Rain water harvested supply (litres per year)} = \text{Rainfall (mm per year)} \times \text{roof area } (\text{m}^2) \times \text{runoff coefficient}$$

The average annual rainfall of a location can be usually obtained from the local meteorological service or one may estimate it by considering the average, minimum and maximum rainfall on a per-month basis. The runoff coefficient is the amount of water that drains from the surface relative to the amount that falls on the surface as rain. For example, a runoff coefficient of 0.7 means that, of the total volume of rain that fell on the catchment surface, 70% drained off the surface and 30% stayed on the catchment area. The material of the roof used gives rise to varying runoff coefficients. Smooth metal sheet catchments with a steep gradient have higher runoff coefficients (0.7 to 0.9) than flat concrete surfaces (0.6 to 0.8). Similarly, tiles have a runoff coefficient between 0.8 to 0.9, brick pavement has a runoff coefficient of around 0.5 to 0.6 and a rocky natural catchment has one of between 0.2 and 0.5.[20]

For better clarity, an example of the calculation based on the rational method is given below:

- Mean annual rainfall = 1500 mm per year
- Roof angle = 70 degrees; sine of the angle = 0.9397
- Roof area = 10 m (length) × 8 m (width) × 0.9397 = 75 $m^2$
- Roof surface is smooth corrugated metal. This surface is assumed to have a runoff coefficient of 0.8.

Supply = 1500 × 75 × 0.8 = 90 000 liters (approx. 19 800 gallons) per year.

### 3.5.2 Gutter

Guttering is used to transport rain water from the roof to the storage vessel. It is usually fixed to the building just below the roof and collects water as it falls from the roof. Gutters are available in a variety of materials, shapes and forms such as poly vinyl chloride (PVC), asbestos, galvanized iron[9] or any other inert substance, as the pH of rain water can be low (acidic) and may cause corrosion and mobilization of metals in metal pipes. PVC gutters are recommended since they do not rust and water quality will be maintained over a long period. In developing countries, where factory made gutters are an expensive option, gutters are generally made from low cost locally available materials.

For instance, V-shaped gutters from a galvanized steel sheet can be easily made by cutting and folding the flat galvanized sheet with the help of simple tools that are already available to the locals.[11] The correct installation of gutters is most important.[19] The size of the gutter should be decided based on the flow during the highest intensity of rainfall.[11] Gutters should not have flat areas where debris can collect and water may pool as these could result in breeding grounds for mosquitoes. They must generally slope towards the storage tank. Maintenance of the gutters is of paramount importance to reduce the collection of debris, leaf litter *etc.*[12]

### 3.5.3 Storage Vessel

Storage tanks for collecting harvested rain water may be located either above or below the ground and may be built either as a part of the building or as a separate unit situated some distance away from the building. Storage tanks built above the ground allow for easy inspection for leakages, extraction of water by gravity using taps and they can be easily raised above the ground level to increase the water pressure. On the other hand, these tanks require more space, are relatively expensive and more easily affected and damaged by weather conditions. The water harvesting surface also needs to be at a higher level, else again, pumping costs are involved. The advantages of underground tanks are manifold. They require little or no space above ground, they are not obstructive, the stored water is cooler and they are less expensive.

However, on the flip side, they are more difficult to empty for cleaning, leaks and failures are difficult to detect, there is possible contamination from groundwater, there is potential danger of falling into the tank if left open and heavy vehicles driven over underground storage tanks can cause damage to the structure. Therefore, careful decision on employing above or below ground storage tanks is imperative. An important aspect in the building of the tanks is their material of construction.

There are myriad types of tanks available that are made of a variety of construction materials such as concrete, fibreglass, polyethylene, stainless steel *etc.* The main requirement is that the material of construction should be inert. Since open storage tanks can lead to contamination from insects, debris and breeding mosquitoes, a solid and secure cover for the tank is recommended. Periodic cleaning of the tank with chlorine solution is generally carried out, especially if the harvested water is intended for potable use and must be stored over an extended period. To maximise the efficiency, storage tanks should be located as close to both the supply and demand locations as possible. They should be placed at an elevated level to facilitate the use of gravity or lower stress on pumps. It is desirable to have the tanks situated away from any sources of pollution such as sewage systems. Additional devices such as an overflow pipe, an indicator of the amount of water in the storage tank, a vent for air circulation and a system for protection against rodents and insects may be also required.[7]

An area of supreme importance is the calculation of the area of the storage tank as the storage tank usually represents the biggest capital investment element of a RWH system. Hence, it needs careful design to provide optimal storage capacity while keeping the cost at a minimum. It must be designed to meet the water requirements of the household, rainfall and catchment availability. For example, if we consider the sample calculation described in section 3.5.1, the potential rain water harvested in a year is 90 000 litres. However, the storage tank capacity must be designed based on the dry period when there is little or no rainfall. If we assume this period to last for 120 days (this information can be obtained from the local meteorological department), then the drinking water requirement during the dry period, for a household of 5 members with a usage of 10 litres of water per person will be as follows:

$$\text{Drinking water requirement for the family in a dry period (litres)}$$
$$= 120 \text{ days} \times 5 \text{ members} \times 10 \text{ litres} = 6000 \text{ litres}$$

Generally, the tank is built 20% larger than required as a safety measure. Therefore, in the above case, a storage tank of capacity 7200 litres will be required.

Approaches like these used for the design of equalization vessels or in effluent treatment systems can be adapted in this case as well.[11]

### 3.5.4 First Flush Device

When rain water gets collected on the catchment area, it often gets contaminated with unwanted material such as animal and bird droppings, debris and dust particles, all of which can pollute the stored water should this first flush enter the storage tank. This can be more pronounced if the rainfall occurs after long dry spells. A first flush device is used to divert this water away from the storage tank so the quality of stored water is not affected. It requires simple installation, forms part of a downpipe and works by channeling the first flow of rain water down the downpipe to its base where it encounters a cap with a small drain hole. This allows the first flow of rain water with the debris to settle at the base of the downpipe with cleaner water settling on the top. This permits relatively cleaner water to enter the storage tank. Figure 3.2 shows the basic design of a first flush system.

The basic design is enhanced by using a ball valve that floats and sits on top of the water column. This ball valve separates the dirty first flush from the cleaner water once the water column in the downpipe floats the ball valve to the constriction in the neck of the downpipe.[12] For maintenance, it is important to ensure that the cap of the first flush device is easily removable to facilitate cleaning. Periodic inspection of the pipe to ensure that it is not clogged with debris and that it drains properly is also necessary. The amount of water that needs to be diverted as the first flush and the length of the downpipe that is required can be calculated as follows:[12]

**Known data:**
Roof length = 9 m
Roof width = 6 m
Pipe diameter = 160 mm. Therefore, radius = 80 mm.

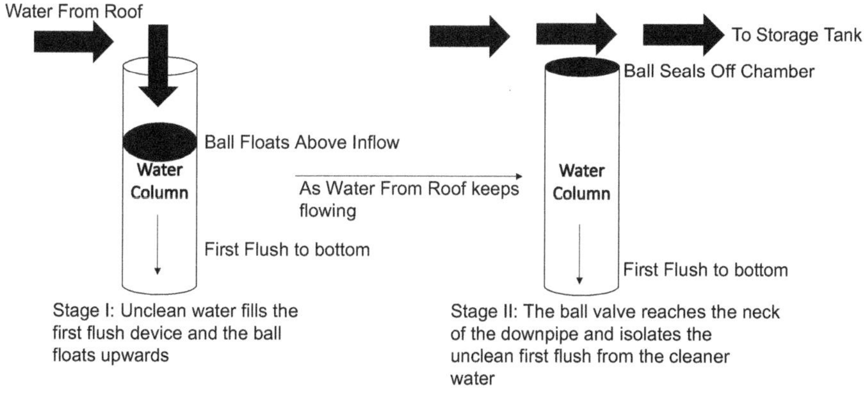


**Figure 3.2** First flush device.

It is assumed that a depth of 0.5 mm of rainfall on the roof is needed to wash off the collected pollutants.

$$\textbf{Volume of diverted water (litre)} = \text{Roof length (m)} \times \text{roof width (m)} \times 0.5\ \text{mm}$$
$$= 9 \times 6 \times 0.5 = 27\ \text{litres}$$

$$\textbf{Pipe length (m)} = \text{Volume of diverted water in litres}$$
$$\div\ \text{cross sectional areaof the pipe}\ (\pi r^2)$$
$$= 27 \div [3.14 \times (80)^2 \times 0.001] = 1.34\ \text{m}$$

### 3.5.5 Quality of Harvested Rain Water and Health

Harvested rain water quality is of paramount importance, especially if it is employed for potable purposes. The harvested rain water can become contaminated by nearby sewage pipes and from unclean catchment areas. Many varieties of chemical contaminants have been found in harvested rain water such as heavy metals, herbicides, pesticides and polycyclic aromatic hydrocarbons. Additionally, several reports are present in the literature that indicate that harvested rain water may pose a public health risk due to the presence of microbial pathogens. Several microorganisms like total coliforms, *E. coli*, Fecal coliforms, *Salmonella* sp., and *Giardia lamblia* have been found in roof runoff water. Guidelines for rain water use suggest that total coliforms and *E. coli* are not detectable at counts <1 CFU per 100 ml[13] and the WHO state that the total coliforms for a water resource should be <10 CFU per 100 ml in 95% of samples taken. When the levels are 20 CFU per 100 ml, it recommends further treatment for the drinking water.[13]

Rainwater tanks should be designed to protect the water from contamination by leaves, dust, insects, vermin and other industrial or agricultural pollutants. Tanks should be built away from trees, they should have good fitting closures and be maintained in good condition. First flush devices are used to entrap the contaminates and prevent their entry into the storage tank, as discussed in the previous sections. Water that is relatively clean at entry will usually improve further in quality if allowed to settle for some period. Microorganisms entering the tank will not survive if the water is clean. Algal growth in tanks can be prevented by building them in cooler areas away from sunlight. Fencing around the area of RWH will prevent animals from fouling the area and be a safety measure for children playing around the tank. Water collected around the tanks should be drained immediately to prevent water stagnation and breeding of mosquitoes. A mosquito proof screen on the openings of the tank should be good enough to prevent insect vectors.[7]

## 3.6 Treatment of Harvested Rain Water

After taking ample precautions, it is natural that some pollutants may find their way into the storage tank. There are literature reports that point out that rain water collected from rooftops and directly from air has been proven to be microbiologically contaminated to a large extent, which includes pathogenic coliforms and fecal streptococci. These studies reveal that harvested rain water cannot be used before appropriate treatment.

Therefore, treating the harvested rain water suitably is very essential, especially if it is intended for potable purposes. There are several techniques that are available, some of which are established methods such as boiling water, addition of chlorine and slow sand filtration, and others are more recent such as SODIS *etc.* A survey conducted by the American Rainwater Catchment Systems Association revealed that UV irradiation and chlorination are the most common techniques employed to disinfect rain water harvested for drinking purposes.[14] Table 3.2 enlists some of the techniques for disinfection of harvested rainwater and the following section describes them in detail.

### 3.6.1 Chemical Disinfection

Adding chlorine is an established and conventional method of treating harvested rain water just like treating any other type of raw water. Chlorine has been used for drinking water disinfection since the beginning of the last century and it remains a choice globally. It is commonly employed for disinfection due to its ability to deactivate most microorganisms and because chlorine is relatively cheap and easily available. It is highly effective against nearly all the water-borne pathogens excepting some viruses and cysts such as *C. parvum* oocycts and the Mycobacteria species. The problem of by-product formation in chlorine disinfection emerged in the mid-1970s with the advent of gas chromatography.

Chlorine has an inherent tendency to react with the organic matter that may be present at the bottom of rain water storage tanks to produce undesirable by-products. Therefore, it is often applied after removal of the water from the tanks. Chlorine contributes to taste and odour related problems and chlorine gas storage poses a major health risk and therefore may not always be a method of choice. An alternative chemical disinfectant is silver as its antimicrobial activity has been well known since ancient times. Silver has been approved by both the EPA and WHO as safe for human consumption. Moreover, in water, silver concentrations needed for microbial disinfection do not impart taste, odour or colour to water.

It is interesting to note that a lifetime total intake of 10 g of silver can be considered to have no observable adverse effects for humans. Although silver has been used for water disinfection, it was only a few years back that it was considered for disinfection of harvested rain water. In an interesting study conducted by Nawaz *et al.* in 2012, silver was employed as a disinfectant

**Table 3.2** Techniques for treating harvested rain water.

| S. no. | Pollutant | Treatment | Remarks | Reference |
|---|---|---|---|---|
| 1. | Microorganisms | Chlorination | Should meet the level of 0.4 to 0.5 mg $L^{-1}$ free chlorine in the form of chlorine gas or tablets. | 26 |
| 2. | Bacteria and Protozoa | Slow sand filtration | Most bacteria (81–100% removal)<br>Protozoa (91.98–100% removal) | 16 |
| 3. | Particulate matter | Rapid sand filtration | The water flows through sand layers covered with activated charcoal or anthracite layers. These are usually depth filters where the small particles are trapped in the pores on their way through the filter rather than the surface, which can result in clogging. | 27 |
| 4. | Microorganisms | SODIS | Carried out in plastic bottles in batch mode. Solar reactors can produce more (100 L) disinfected water per square meter of the reactor per day. Not suitable when TDS is above 10 mg $l^{-1}$. | 20 |
| 5. | Microorganisms and Particulate matter | Metal membrane filters with ozonation and aeration | Metal membrane filters of size 1 to 5 micrometer are usually good enough to remove microorganisms as well as particles. Ozonation and aeration on the feed side helps prevent membrane clogging and inactivate microorganisms. | 28 |
| 6. | Turbidity, organic matter and Bacteria | Filtration/adsorption and UV | Turbidity (98% removal). Organic matter (38–49% removal).<br>Bacteria (total coliforms – 99% removal) | 29 |
| 7. | *Pseudomonas aeruginosa* and *E. coli* | Silver disinfection | Dosage of 0.01 to 0.1 mg $l^{-1}$ silver was studied. Optimum dosage of 0.08 mg $l^{-1}$ has been reported for 95–99% disinfection of both microorganisms. | 30 |
| 8. | Particulate matter | Up flow filtration and back flow back flushing.<br>A prototype was tested using 3 different filter fabrics that varied in material and specifications | Using a polypropylene woven filter fabric with 75 mm pore size and 1.07 mm thickness, the prototype system operated with 68% treatment efficiency and 100% backwash efficiency. | 17 |
| 9. | Microorganisms | SOCODIS and SOCODIS with natural ingredients like lemon and vinegar | The efficiency of SODIS is increased by using solar collector disinfection (SOCODIS), which works by concentrating the sunlight. Further enhancement of disinfection was obtained by lowering the pH to 3 with the help of commonly available food preservatives such as lemon (0.25%) or vinegar (0.17%). | 31 |
| 10. | Microorganisms | Boiling | Temperature range from 55 to 65 °C was critical for effective elimination of enteric/pathogenic bacterial components. It has been recommended that hot water systems should operate at a minimum of 60 °C. | 32 |

against *E.coli* and *P. aeroginosa* in harvested rain water. It was observed that at higher concentrations of silver (0.08–0.1 $mg\,l^{-1}$), both the microbes were inactivated in about 10 h with a high residual effect of silver as compared to that at lower concentrations (0.01–0.01 $mg\,l^{-1}$), which required a longer time of 14 to 24 h. They reported that *E. coli* was more sensitive to silver as compared to *P. aeroginosa*.[15]

### 3.6.2 Filtration

Among the filtration methods commonly used for water treatment, slow sand filtration appears to be a method of choice for treating harvested rain water mainly due to its cost effectiveness. It essentially consists of graded sand with the finest below and the coarsest on the top. Sand filtration works due to the formation of a thin biofilm on top of the filter and its efficiency depends on a constant flow of water through it. Although it successfully removes most bacteria (81–100% removal) and protozoa (91.98–100% removal), it is not effective for viruses.[16] It is the most common type of treatment unit employed in developing countries. For the removal of fine suspended particles, a wide spectrum of filters such as membranes, sand, granular activated carbon (GAC) and expanded clay and gravel are available.

Modern designs such as a self-cleaning filtration system for harvested rain water treatment were investigated by Vieira *et al.* in 2013. The unit essentially consisted of an up-flow filtration and down flow backwashing mechanism. This system operated with 68% treatment efficiency and 100% backwash efficiency. Thus, the self-cleaning feature of this device improved the performance of the filter and maintenance requirements were also reduced. Moreover, the compact footprint, no energy usage and easy installation in local areas were reported as additional merits by the authors.[17]

Another way to enhance the filtration efficiency and disinfection of harvested rain water is to employ hybrid techniques. These methods usually combine filtration with adsorption and/or UV irradiation. In one such hybrid method, activated carbon filter in the form of pressed blocks was used either as such or precoated with silver. Water passing through this filter was subjected to UV disinfection. Such systems had an operating life of only 6–8 months and were expensive.

Yet another alternative was studied by Naddeo *et al.* in 2013. An innovative treatment unit called FAD (Filtration Adsorption Disinfection) was used to treat harvested rain water for potable use. Water was passed through a filter followed by adsorption on GAC (Granulated Activated Carbon) and then subjected to UV disinfection. Turbidity (98% removal) and organic matter content (38–49% removal) were effectively reduced by this novel concept.[18] However, it has been observed that deficiencies in these hybrid processes and/or deterioration of the water quality in the distribution system due to the absence of a disinfectant residual often led to a less than expected bacterial concentration in the harvested rain water.

### 3.6.3 Ultraviolet Irradiation

Irradiation with UV systems is yet another method of disinfection that has been used as it is easy to install and economical to operate and maintain. In recent years, there has been a renewed interest in UV systems for water disinfection due to their ability to inactivate (>99.9%) chlorine resistant protozoans, *Giardia lamblia* cysts and *Cryptosporidium parvum* oocycts. Generally, a 16 mJ dose of UV is sufficient for disinfection of non-pathogenic organisms. However, for inactivation of *Giardia* and *Cryptosporidium*, a minimum dose of 40 mJ is required. However, for UV systems to function, water must have low turbidity to enable UV penetration and microbial de-activation.[18] Therefore, these lamps need periodic cleaning, especially if placed in a submerged configuration. Due to their finite life span, they must be periodically replaced. Further research on the use of UV for treating harvested rain water has led scientists to investigate the ability of UV to treat micropollutants in rain water.

It has been reported that a Vacuum–UV (VUV) process is a promising advanced treatment of rain water due to its high efficiency, easy installation and safe rain water reuse. VUV is an emerging technique and is based on advance oxidation processes (AOPs). It works by generating $HO^{\bullet}$ by photo-ionization of water molecules after absorbing photons having short wave-lengths, *e.g.*, 185 nm. Most importantly, this process does not require the addition of extra oxidants and thus their subsequent removal. Li *et al.*, in 2015, reported the first such study where six micropollutants including phenols and recalcitrant aromatics in harvested rain water were successfully treated with VUV.[19]

### 3.6.4 Solar Disinfection

Among the myriad techniques of treating harvested rain water, solar disinfection (SODIS) is an economical, sustainable and simple method to disinfect potable water, especially in developing countries, and especially as sun is a free natural source of energy and is plentiful in most developing countries worldwide. SODIS can be carried out on a batch scale that requires simple Polyethylene terephthalate (PET) bottles. These are easily available and are considered safe as no migration of photoproducts or additives from PET bottles into water was observed. SODIS may also be carried out in continuous flow reactors, which can produce approximately 100 L of treated water per sq. of solar collector per day.[20] Over a period, attempts were made to enhance the efficiency of SODIS.

Various approaches have been investigated such as using different backing surfaces and solar concentrators. Another approach has been tried that involved the use of titanium dioxide as a catalyst either in suspended or immobilized form to enhance SODIS. This was useful in the reduction of fecal coliforms and *E. coli.* Compared to *E. coli*, 2.4 times as much solar energy was needed to inactivate *Candid* sp., *Geotrichum* sp., and spores of

*Aspergillus flavus* and about 6.4 times as much solar energy was needed to inactivate *Penicillium* spores. Yet another innovative technique to accelerate the efficiency of SODIS is by using low cost additives such as hydrogen peroxide, copper metal and lime juice.[20] Another approach was the use of a solar collector disinfection (SOCO-DIS) system where an almost 30 to 40% increase in disinfection was observed in harvested rain water.

In a very recent study, a comparative analysis of solar pasteurization (SOPAS) *versus* solar disinfection (SODIS) for harvested rain water was carried out. The main aim was to compare SOPAS (70 to 79 °C; 80 to 89 °C; and ≥90 °C) to SODIS (6 and 8 h) for their efficiency in improving the microbial and chemical quality of harvested rain water. Both SOPAS and SODIS reduced the count of indicator microorganisms to below the detection limit (>99%). SOPAS yielded a 2 and 3 log reduction in viable *Legionella* and *Pseudomonas* counts, respectively, and SODIS after 8 hours resulted in a 2 log and 1 log reduction in viable *Legionella* and *Pseudomonas* counts, respectively, and could be considered as a cost-effective treatment technique for harvested rain water.[21] Although several studies have highlighted the potential of solar disinfection for rain water treatment, it is important to note that when the concentration of suspended solids exceeds 10 $mg\,L^{-1}$, other means to eliminate the particulate matter such as filtration become imperative.

## 3.7 Merits and Demerits

The advantages of rain water harvesting are manifold. It relieves demand and decreases the dependency on underground and surface water sources. It is cost effective as it reduces water bills and running costs are low. It is a simple yet flexible technology and local people can be easily trained to build, operate and maintain a RWH system. Moreover, water can be delivered directly to the household, thereby relieving the women and children of the burden of carrying water, saving time and energy.[22] On the other hand, there are some demerits to rain water harvesting such as dependency on climatic patterns, storage capacity limitations and contamination from poor collection and storage methods. However, several attempts in rain water harvesting have been successfully made in several developing countries and the following section describes some of these interesting stories from India.

## 3.8 Some Attempts in Rain Water Harvesting in India

India as a leading developing country is reeling under a continued water crisis that shows no signs of abating. Excessive demand coupled with mismanaged water resources, erratic weather patterns and climate change has led to a severe water shortage in parts of the country. To put it simply, there's too much demand and not enough supply. India's burgeoning population woes aside, climate change is seen as a significant factor in the country's water crisis. Scorching summers and shortened winters have resulted in reduced snow cover and retreating glaciers in the Himalayas—the ice melt from these areas is what feeds the country's northern rivers. Add to that a fickle monsoon that threaten to disrupt long-established crop patterns and you have a recipe for disaster in many parts of the country. Then there's agriculture, which, by some estimates, accounts for 75% of India's water consumption. Farmers are sometimes slow to change what they sow and continue to grow thirsty crops like rice and sugarcane in parched regions. They meet these irrigation demands by indiscriminately extracting groundwater. This has led to a substantial drop in the levels of India's water table, which is another major concern. Though the government has undertaken various large and expensive projects such as irrigation, river linking, dams, canals, *etc.* to check the water crisis, these projects are not enough or sufficiently sustainable to cater to the ever-growing demand for water.

Meanwhile, the water shortage has reached such a level that even the highest rainfall receiver, Cherrapunji in Meghalaya, has started to face water scarcity in summer. Earlier water shortages used to be seasonal, mostly restricted to summer. Now, this problem is haunting the country even during the monsoon. With such a bleak picture, instead of relying entirely on such grand schemes, the time has come to find other complementary solutions. If we look at the water crisis in India, we can easily see that more water is being utilized than that being stored. The only way to reduce this water deficit is rain water harvesting, particularly, *in-situ* rain water harvesting. It is a perfect inexpensive solution that can be successfully implemented through small, localised and focused initiatives of individuals and communities. Plenty of such efforts have worked wonders, but they have failed to garner attention. In the following, such stories that perfectly demonstrate the efficacy of *in-situ* rain water harvesting methods are described.[38,39]

### 3.8.1 Work Initiated by Dr Pol (Satara, Maharashtra)

A local public cleaning drive at Ajinkyatara Fort, Satara, in the state of Maharashtra in India, which later turned into a soil and water conservation project, has now become an excellent *in-situ* rain water harvesting model being replicated in nearby villages. A cleaning campaign of the historical 16th century Ajinkyatara Fort, Satara was organised to which only a handful of people volunteered. Dr Pol was one of the few volunteers. After cleaning the fort, Dr Pol suggested utilising hundreds of acres of fort area to harvest soil and water. Considering it to be a likely solution to their water woes, many volunteers from the town joined Dr Pol. Led by Dr Pol, volunteers dug hundreds of continuous covered trenches (CCTs), loose boulder structures (LBSs) and small check dams (Figure 3.3). They planted and nurtured thousands of trees. The fruits of the community labour became evident when water levels increased from 200 feet to 40–50 feet near the foothills.

With this success, Dr Pol went on spreading awareness of the importance of water harvesting and 'shramdaan' or voluntary labour in service of the community. They also showed villagers how to connect with the government

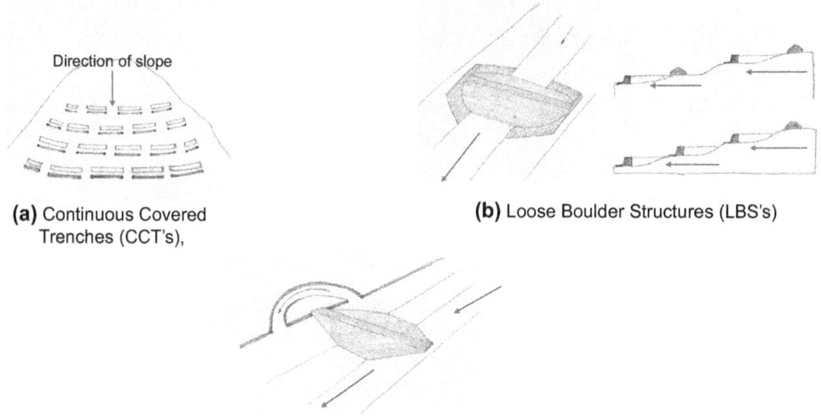


**(a)** Continuous Covered Trenches (CCT's),

**(b)** Loose Boulder Structures (LBS's)

**(c)** Small Check Dams

**Figure 3.3** Attempts at rain water harvesting by Dr Pol.

and access funds for the same. Many renowned NGOs like the Paani and Naam foundation supported Dr Pol to expand this mission to various villages. One dramatic example worth mentioning here is that of Velu village. People from the water-deprived village approached Dr Pol. Volunteers from the village carried out the work suggested by Dr Pol. A mere 4 cm of rain helped to raise the water level of wells and ponds. Experiencing the importance of water harvesting, they even went on to interlink the percolation ponds to avoid the spilling of excess water.

As a result of these efforts, the villages became water self-sufficient. Cultivation resumed, and farmers started using drip irrigation to cultivate potatoes, ginger, chillies, *etc.* With the availability of fodder, dwindled livestock began to flourish, and hence, milk production increased. Further, with the revival of income, migration in search of jobs reduced.[40]

## 3.8.2 Work Initiated by Lakshman Singh (Jaipur)

Lapodiya, a village on the Jaipur–Ajmer highway, in the state of Rajasthan, India, literally means 'a fool's village' and the people lived up to it. The people were involved in fighting, messing with each other's farms, *etc.* and they ended up being poor. Further, the lack of water – both due to deficient rainfall and neglect of traditional water bodies like tanks, wells, and ponds, aggravated their woes. Lakshman Singh, a class 7 dropout resident of the village, saw the scarcity of water as a primary cause of the village's problems. He believed that if the water shortage could be reduced, the village would prosper in harmony. This led to the innovation of a new water harvesting technique known as a 'Chauka system.' The journey of restoration of the ponds and wells was initially faced with apathy from the villagers due to the unavailability of monetary funds. Instead of feeling discouraged, he started working alone. A few days later, many hands joined him, solving the problem of a lack of money through voluntary labour. Percolation ponds, two tanks, Phool Sagar and Dev Sagar, and one irrigation tank, Anna Sagar, were dug. As the tanks began holding water, they observed a rise in open well water level. This convinced the villagers of the importance of water harvesting. Now, all they had to do was to collect more water to ensure all year water security. As water couldn't be collected in farm fields, it was decided to utilize pasture lands. But, water from pasture lands used to run down into a Nala or drain where it was wasted. This water was needed to be collected and diverted to the existing tanks. But, traditional contour trenches and bunds were not suitable as water ultimately ended up in a Nala or drain.

After studying the topography of Lapodiya, Lakshman Singh came up with a unique system of Chauka. The Chauka is a rectangular enclosure surrounded by soil ridges on three sides and that is open on one side. A series of such Chaukas was constructed, creating a water network (Figure 3.4). This series of Chaukas fill gradually with rain water. The excess rain water moves sideways from one Chauka to the next Chauka. Through a common waterway, excess water is diverted to a tank instead of a Nala or drain. The Chauka system allowed pasture land to retain water, helping groundwater recharge as well as reducing soil erosion. Further, it allowed villagers to decide how much water needs to be collected by raising or lowering the height of the ridge. Depending on the level of retained water, the spread of unwanted varieties of grass can be controlled. For example, thorny and unpalatable grasses can be reduced when the retained water level is more than nine inches. This ensures good quality of fodder for grazing animals.

Lakshman Singh along with his local group, Gramin Vikas Navyuvak Mandal Laporiya (GVMNL), went on to implement the Lapodiya model in neighbouring villages. The GVNML also developed Chauka kits for easier understanding and trained villagers for pasture revival. They also improved the stock of cattle by crossbreeding with Gir cattle. As a result of this, Laporiya now produces 1500 litres of milk a day. This exercise has also changed the perspective of the villagers towards sustainable development. They have now started to take care of their cattle, birds, and trees as well.[41,42]

### 3.8.3 Work Initiated by Prof A. G. Chachadi (Goa)

The university campus of Goa University is located on the periphery of Panaji, which is also the capital city of Goa, a small state on the Western Ghats of India. It stretches across 173 hectares of the area on the Taleigao plateau overlooking the Zuari River, which enters the Arabian Sea. The area has hard laterite rock of varying thickness underneath, which contains different layers of clay and fractured and withered rocks, which can be used as confined aquifers. The climate is pleasant in Goa, and minimal fluctuations are seen in the temperature. The maximum temperature achieved is about 35 °C. The rainfall starts in June and lasts till September. 400 cm rainfall is

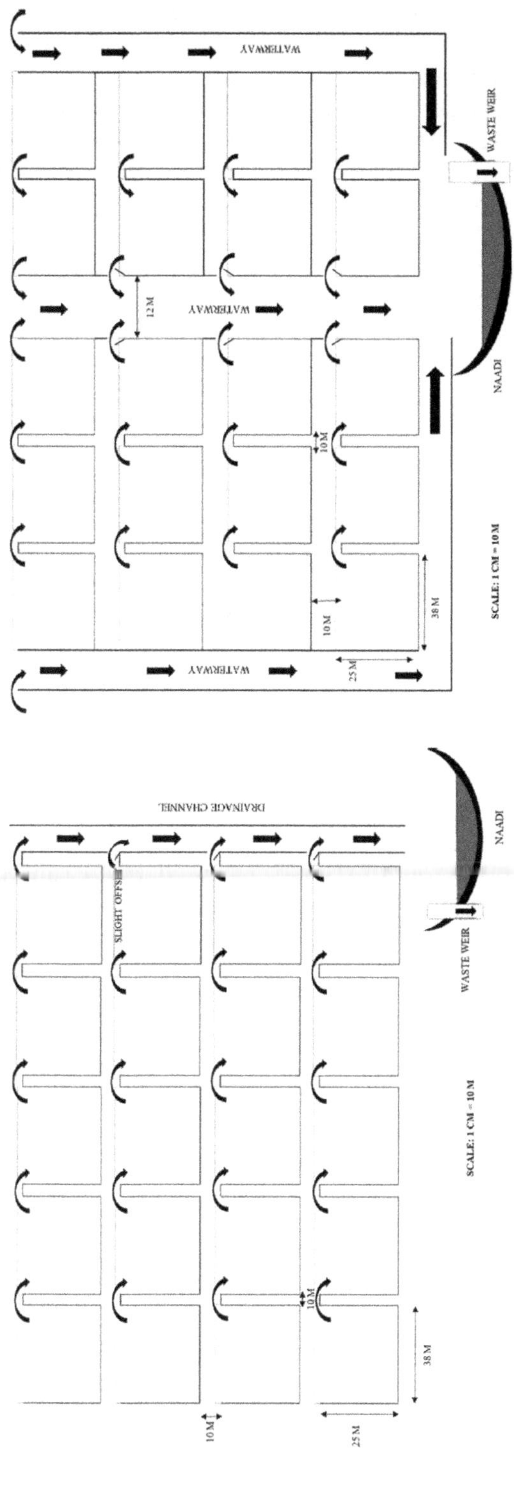


**Figure 3.4** Models of the Chauka system: (a) simple model; (b) model with middle waterway.

recorded in vegetated areas whereas 250 cm rainfall is recorded in the coastal region annually. Overall, an average of 320 cm per annum rainfall is recorded in Goa, making it lie in the heavy rainfall zone. The campus of the university was mostly dependent on the groundwater obtained from the 12 borewells on campus and the supply of water from local governing bodies. During summer, the local supplies would be cut by almost 50%, which led to overuse of the groundwater to fulfill the needs of 1500 people living on the campus. Some of the borewells dried up due to this. In 2007, Rain Water Harvesting was constructed under the guidance of Prof A. G. Chachadi, Department of Earth Sciences, Goa University, to recharge the fast depleting aquifers.

The system currently has two main structures, one main structure for harvesting surface run-off with a catchment area of 1.5 hectare and a second for the collection of run-off from uplifted and built up areas on campus. To maintain the cleanliness of the water, a rubble wall was built around this natural depression, *i.e.*, the 1.5 hectare structure. A deep trench (dimensions: 20 m×10 m and 2 m) was built at the centre of the pond, and a retention concrete wall of 0.5 m height was built around this trench to prevent direct entry of the pond water. The runoff water passes into the deep trench ensuring that the suspended particles settle down before water enters into the sand filter (dimensions: 3 m×3 m×3 m). Finally, after getting through the sand filter, the water enters the main recharge well trench (dimensions: 3 m×3 m×4 m). The deep borewell is perforated both at the surface as well as at the bottom where it is exposed to the aquifer. The runoff water enters through these perforations wrapped with nylon cloth and directly enters into the aquifers at deeper levels. The borewell is drilled to a depth of 89 m below ground and taps the aquifer at 65 m depth. In the well, water enters through the top 5 m length of the slotted casing pipe. Inside the bore, water is released to the aquifer between 65 m and 89 m depth through a slotted casing pipe. Between 89 m and 65 m depth range, the borewell is cased with a blank pipe as the geological layer in this depth range is made up of clay (Figure 3.5).

The situation changed after the implementation of the scheme. The main structure in the campus collected a total of 55.5 million litres of water out of which around 38 million litres were consumed in the recharging of the aquifers. The demand of 39 million litres in 2010 was met from this water. The secondary structure has an area of 400 sq m, which harvested 0.26, 0.18 and 0.9 million litres of water in the years from 2008 to 2010. During May 2009, it was found that 13 to 15% of the groundwater was recovered. This was achieved within a year of operation of the plant.

The benefits of this programme were received at only a minimal maintenance cost of INR 8000 to 10 000 (≃100 USD) per year. The capital costs were recovered within 5 years for the main structure and around 10 to 11 years for the secondary structures. By calculating the cut in electricity bills, water bills, and tanker costs, it was estimated that this RWH system saves up to the sum of INR 2.8 million annually for the university.[43]

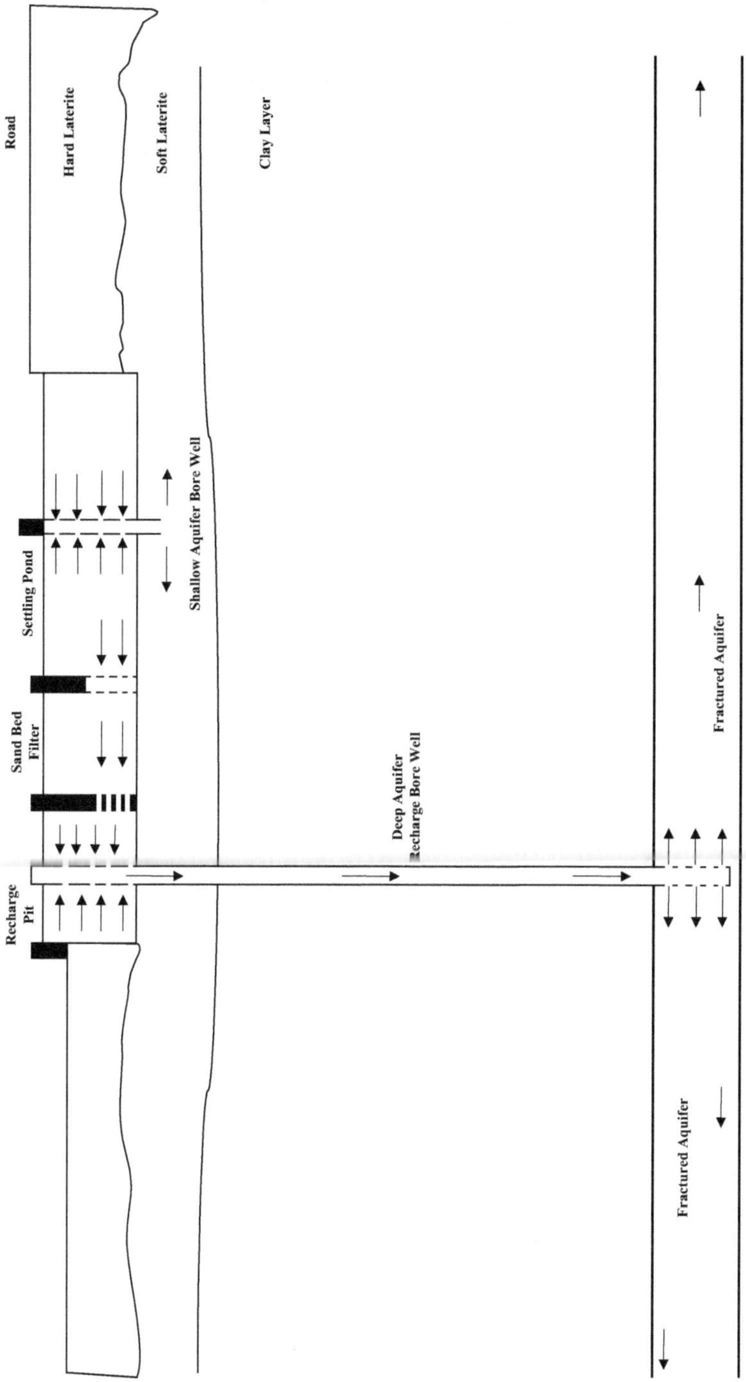


**Figure 3.5** Surface water run-off system of the main structure.

### 3.8.4 Work Initiated by a Farmers Community (Mewat)

This village falls in the scarce rainfall region and has one of the lowest development indices in the country. Mewat lies at the foothills of the Aravalli ranges, in the Northern part of Rajasthan, a state of India. Thus, there is a huge catchment area uphill. Due to there being no obstruction in between, the water flows furiously through the village. The percolation rate is too low due to the high gradient. The high velocity also caused agricultural destruction and soil erosion. There was only a 200 meter wide stretch under the hills where fresh groundwater was available; otherwise, the rest of the village had saline groundwater. Farmers of this village used to grow rain-fed mustard and tomato only if someone had a tube well within these 200 meters. The population mainly depended on groundwater, and it was depleting at a rate of 300 to 500 mm per year. Thus, the ridge and valley system was adopted to store and percolate the same flowing water. The ridge to valley approach was unique as it was able to reduce the flow velocity and minimize the chances of erosion in the agricultural fields and silting in storage basins, helping to increase the recharging potential. This was carried out through a series of check dams wherever possible with the idea that backwaters of a check dam touch the preceding check dam in the stream such that the whole length of this stream becomes a storage and percolation basin. After facing hardships in achieving community contributions, finally, the two check dams were completed in June 2007. That year, the rainfall was not very great, still, an average rise of 7 feet was noted in the water level of open wells. It is estimated that annually, around 507 million litres of water are harvested through an area of 26 000 sq m. Many farmers have also confirmed that the wells are now supplying sweet water as compared to saline water before the construction.[44]

## 3.9 Conclusions

Rain water harvesting is a very effective method for water conservation and is a promising solution to address the water demands globally. For developing countries, RWH appears to be a simple, economical way to provide water for domestic needs. It is apparent from the preceding sections that the quality of the rain water harvested changes in purity as it encounters various pollutants from air emissions, catchment surfaces, storage tanks *etc.* Thus, treating the harvested rain water before consumption is mandatory to ensure the safety and health of the end user.

Among various methods that are discussed here, filtration, SODIS, UV and chemical disinfection and a combination of these are still preferred as treatment methods for obtaining potable water from rain water. Rainwater, nature's gift to humankind, will always remain precious to all living beings on this planet. It is also our responsibility to effectively collect, treat, manage and supply this valuable resource so that demands of drinking water are met, especially in the developing countries.

## 3.10 Further Research and Recommendations

Listed below are a few areas where further research needs to be conducted:

- More studies are needed to enhance the design of the RWH unit, thereby improving its efficiency, *e.g.* novel roofing materials that do not contribute to polluting the rain water, better designing of storage cisterns and piping to ensure a safe and good supply of water.
- Novel treatment techniques that are cheaper than the existing ones and more effective in the removal of chemical, physical and biological contaminants are the need of the hour.
- There should be more emphasis on the economics of the RWH systems in developing countries. Such studies are scarce in the literature. The Appendix lists some information related to the cost of typical RWH systems.
- Further investigation to reduce evaporation of harvested rain water, to prevent mosquito breeding and reduce the animal impacts is also required.

# Appendix

## Costing of Rain Water Harvesting System

Table 3.3 contains information related to the cost estimation of typical rain water harvesting installations. This table enlists various component, maintenance and labour costs for RWH systems from different parts of the world as reported in the literature.

**Table 3.3** Cost of rain water harvesting systems.

| S. no. | Component of RWH system/maintenance activity | Salient features | Approximate cost | Reference |
|---|---|---|---|---|
| 1. | Passive RWH system in households that uses simple plastic rain barrels | Capacity of around 50 gallons. Cost depends on size, quality and manufacturer. | \$70 | 33 |
| | | | US\$0.53 per gallon to over US\$0.74 per gallon | 12 |
| | | | £70–80 for a 200 litre container to around £350 for a 1000 litre system. | 10 |
| 2. | Active RWH systems use Storage Cisterns or Tanks | Range in price depending on the material, size, and shape. Large systems have additional costs due to pumps, filters and treatment *etc.* | Simple systems: between \$1.50 and \$3.00 per gallon of storage<br>Large complex systems: an additional \$2–\$5 per gallon of harvesting system capacity. | 33 |

Table 3.3 (*Continued*)

| S. no. | Component of RWH system/maintenance activity | Salient features | Approximate cost | Reference |
|---|---|---|---|---|
| 3. | Routine maintenance activities for typical cistern systems | Inspection of the cistern periodically, roof and inflow filter cleaning | \$ 260 per year<br>\$ 480 per year | 34 |
| 4. | Infrequent maintenance costs | Pump replacement<br>System flushing and sediment removal from tanks | \$ 198 per year<br>\$ 130 per year | 34 |
| 5. | Clean water tank | 150 $m^3$ capacity | 350 CNY per cubic Meters (cost including direct and other expenses) | 35 |
| 6. | Ferrocement tanks with skeletal cage | Total capacity of tank – 10 000 litres | INR 15 800 | 36 |
| 7. | PVC piping for rain water pipes | Pipe diameter size<br>– 110 mm diameter<br>– 200 mm diameter | <br>INR 165<br>INR 275 | 36 |
| 8. | Excavation requirements for RWH system installation | For rocky foundations, the installation costs can inflate | US\$1.11 per gallon | 12 |
| 9. | Material of construction, conveyance and labour for building a 5000 l capacity storage tank | Cost includes Precast rings, sand, concrete bags, binding wire, bricks, gate valve, filter media, pipes, first flush *etc.* | INR 6510.00<br>1.30/- per litre is the unit price | 37 |

# References


1. E. Louka, *International Environmental Law: Fairness, Effectiveness, and World Order*, Cambridge University Press, 2006, 536 p. ISBN: 9-780-5-2168-759-1.
2. http://www.sustainable.com.au/rainwater.html (last accessed January 2017).
3. http://www.whollyh2o.org/rainwater-stormwater/item/48-rainwater-harvesting-defined.html (last accessed January 2017).
4. http://www.unep.org/depi/ (last accessed December 2016).
5. http://www.unep.or.jp/ietc/publications/urban/urbanenv-2/index.asp (last accessed January 2017).
6. M. Elliott, A. Armstrong, J. Lobuglio and J. Bartram, *Technologies for Climate Change Adaptation – The Water Sector*, UNEP Risoe Centre, Roskilde, 2011.

7. http://www.practicalaction.org (last accessed November 2016).
8. http://www.uia.org (last accessed November 2016).
9. http://www.unep.org (last accessed December 2016).
10. https://www.renewableenergyhub.co.uk/the-history-of-rainwater-harvesting.html. (last accessed December 2016).
11. http://www.bebuffered.com/downloads/UNHABITAT_Blue_Drop_Series_RWH&Utilisation.pdf (last accessed January 2017).
12. https://www.caribank.org/uploads/2013/08/em-rainwater-handbook-caribbean.pdf (last accessed November 2016).
13. World Health Organization (WHO), *Guideline for Drinking Water Quality*, 3rd edn, World Health Organization, Geneva, Switzerland, 2004.
14. R. B. Thomas, M. J. Kirisits, D. J. Lye and K. A. Kinney, *J. Cleaner Prod.*, 2014, **75**, 166–173.
15. M. Nawaz, M. Y. Han, K. Tschung-il, U. Manzoor and M. T. Amin, *Sci. Total Environ.*, 2012, **431**, 20–25.
16. E. Fewster, A. Mol and C. Wiessent-Brandsma, Proceedings of the 30th WEDC International Conference, Vientiane, Laos PDR, 2004, pp. 1–3.
17. A. Silva Vieira, M. Weeber and E. Ghisi, *Resour., Conserv. Recycl.*, 2013, **78**, 67–73.
18. N. Vincenzo, S. Davide and B. Vincenzo, *J. Hydrol.*, 2013, **498**, 287–291.
19. Z. Li, W. Chen and S. Yue, 5th International Conference on Advanced Engineering Materials and Technology (AEMT 2015).
20. B. Sommer, A. Marino, Y. Solarte, M. L. Salas, C. Dierolf, C. Valiente, D. Mora, R. Rechsteiner, P. Setter, W. Wirojanagud, H. Ajarmeh, A. Al-Hassan and M. Wegelin, *J. Water Supply Res. Technol.*, 1997, **46**, 127–137.
21. A. Strauss, P. H. Dobrowsky, T. Ndlovu, B. Reyneke and W. Khan, *BMC Microbiol.*, 2016, **16**, 289.
22. http://www.eng.warwick.ac.uk/ircsa/ (last accessed January 2017).
23. N. Areerachakul, *J. Sustainable Dev. Energy Water Environ. Syst.*, 2013, **1**(3), 213–226.
24. http://www.gdrc.org (last accessed February 2017).
25. https://en.wikipedia.org/wiki/Rainwater_harvesting (last accessed December 2016).
26. G. Palmateer, D. Manz, A. Jurkovic, R. McInnis, S. Unger, K. K. Kwan and B. J. Dutka, *Environ. Toxicol.*, 1999, **14**(2), 217–225.
27. B. Fewreich and H. Horn, *Desalination*, 2010, **251**, 118–124.
28. N. Safapour and R. H. Metcalf, *Appl. Environ. Microbiol.*, 1999, **65**(2), 859–861.
29. V. Naddeo, D. Scannapieco and V. Belgiorno, *J. Hydrol.*, 2013, **498**, 287–291.
30. M. Nawaz, M. Y. Han, T.-I. Kim, U. Manzoor and M. T. Amin, *Sci. Total Environ.*, 2012, **431**, 20–25.
31. M. T. Amin and M. Y. Han, *Desalination*, 2011, **276**(1–3), 416–424.
32. A. Spinks, H. Dunstan, T. Harrison, P. Coombes and G. Kuczera, *Water Res.*, 2006, **40**, 1326–1332.

33. https://www.epa.gov/sites/production/files/2015-11/documents/rainharvesting.pdf (last accessed January 2017).
34. Water Environment Research Foundation (WERF), *BMP and LID Whole Life Cost Models, Version 2.0*, Alexandria, VA. 2009.
35. S. Zhen, Y. Xun, S. Xu and Y. Zhou, Asia-Pacific Energy Equipment Engineering Research Conference (AP3ER 2015), 367–370.
36. http://www.rainwaterharvesting.org/Urban/Costs.htm (last accessed January 2017).
37. http://www.eng.warwick.ac.uk/ircsa/pdf/12th/1/Amarnath-RB.pdf (last accessed January 2017).
38. 'Water in India: Situation and prospects', UNICEF *et al.* 2013.
39. P. S. V. Shankar, *et al.*, India's groundwater challenge and the way forward, *Econ. Polit. Wkly.*, 2011, **46**(2011), 37.
40. ManabiKatoch, 'Meet the Dental Surgeon Who Improved Sanitation and Revived Water Bodies in 70 Maharashtra Villages', The Better India.
41. Mari, 'The Chauka System: An innovative method of water harvesting', Gather and Grow.
42. 'Lakshman Singh - Defeating Drought', UNSUNG-Extraordinary Lives.
43. Interview with Professor AChachadi, Goa University, Goa, January 14, 2014.
44. L. M. Sharma and P. Murada, Harvesting water harnessing life: A case study of Kotla village in Mewat, Rajasthan, 2009.

CHAPTER 4

# *Waste Water Reuse in Developing Countries*

The previous chapter described rain water harvesting and its treatment as one of the methods to conserve and provide safe drinking water to the population in developing nations. Given the severe water crunch across several nations globally, additional water conservation methods to replenish the water bodies are necessary in the years to come. Waste water reuse appears to be yet another prospective methodology that can be successfully implemented in developing countries. Over the last decade, plenty of efforts by experts and stakeholders have resulted in several types of water reuse technologies that can cater to indirect and direct potable use. This chapter deals with waste water reuse with an emphasis on its use in developing countries. Various waste water reuse methods have been described right from the fundamentals of introducing basic waste water terminologies, to waste water treatment guidelines, common pollutants that are encountered in waste waters and the treatment methods involved with suitable examples of implemented projects across the world.

## 4.1 Introduction

The term waste water reuse is usually used synonymously with the terms waste water recycling and waste water reclamation. The U.S. Environmental Protection Agency (USEPA) defines waste water reuse as "using waste water or reclaimed water from one application to another application". The deliberate use of reclaimed water or waste water must follow applicable rules for a beneficial purpose (landscape irrigation, agricultural irrigation, aesthetic uses, ground water recharge, industrial uses and fire protection).[1] The common rationale for waste water reuse is to recognize new water

Drinking Water Treatment for Developing Countries: Physical, Chemical and Biological Pollutants
By Aniruddha Bhalchandra Pandit and Jyoti Kishen Kumar

Published by the Royal Society of Chemistry, www.rsc.org

sources to meet the increasing water demands and to meet progressively higher stringent supply and discharge standards.

Water recycling is characterized as unplanned or planned. When water is drawn from rivers or water bodies that also receive waste water discharges and is reused, treated and piped into the water supply by a variety of natural water harvesting techniques, several times before the end user utilizes the water, then such a case is called unplanned. On the other hand, when projects are specifically designed with the aim of beneficially using recycled water, then such a case is termed as planned recycled water.[2] Water reuse, whether planned or unplanned, is a common practice worldwide and the waste water sources that are used for reclamation come from various origins such as municipal sewage, industrial waste water and storm water runoff.[3,4] These can be used in a spectrum of potable and non-potable reuse applications, as shown in Table 4.1.

Waste water reuse is being practiced worldwide and there are over 3000 water reclamation sites globally where domestic, commercial or industrial waste water is reclaimed. The groundwater replenishment system, Orange county, California, USA, NEWater, Singapore, Torrie/St Andre Water Reclamation Plant, Koksijde, Belgium, and Western Corridor Recycled Water Scheme, Brisbane, Queensland, Australia are some of the global water reuse initiatives.[4]

The latest update on global water reuse was provided by the United Nations World Water Development Report 2017. It is alarming to note that in low income countries, only about 8% of the municipal and industrial waste water they produce undergoes treatment of any kind. In high-income countries, the main reason for waste water treatment and reuse appears to be either to provide an alternative water supply or/and to maintain

**Table 4.1** Categories of waste water reuse.

| S. no. | Category of waste water reuse | Examples of reuse options | Reference |
|---|---|---|---|
| 1. | Urban reuse | Use in school yards, residential landscapes, fire protection, toilet flushing in commercial and industrial buildings and irrigation of public parks | 1 |
| 2. | Recreational reuse | Swimming pools, recreational activities, lakes and ponds recharge | 1, 3 |
| 3. | Environmental reuse | Creating artificial wetlands and enhancing natural wetlands | 1 |
| 4. | Industrial reuse | Process or makeup water, dust separation, power generation and cooling tower water | 1, 3 |
| 5. | Agricultural reuse | Irrigation of nonfood crops, commercial nurseries. High quality reclaimed water is used to irrigate food crops. | 1 |
| 6. | Domestic reuse | Garden use, showering and bathing, toilet flushing and drinking water | 3 |

environmental quality. In developing countries, such practices are scarcely adopted due to a lack of infrastructure, technical and institutional capacity and economic concerns.

However, over the last decade, there have been instances where water reuse initiatives have been successfully implemented, *e.g.* in Windhoek, Namibia and Beaufort West, South Africa (Table 4.3). Simple techniques are available that are effective and economical and particularly suited for developing countries such as the decentralized waste water treatment system (DEWATS) that produce treated waste water of adequate quality levels that can be used for various potential applications like agriculture. Biological treatment processes are also suitable for developing countries given that these processes operate well at higher temperatures and most developing countries have a warm climate. Due to the increasing water demand, use of treated waste water has become a means of enhancing the water supplies in several developing countries. For instance, in the year 2013, 71% of the waste water produced in the Arab States was treated, of which 21% is being used for groundwater recharge and irrigation.[5]

## 4.2 Waste Water Terminology

For the successful implementation of water reuse applications, it is very important that there is a common understanding of the concepts and terminologies used in the planning and execution of these reuse schemes. New concepts and approaches should be taken up in the local and international literature and require clear explanation and standardization. This will benefit the water reuse stakeholders and the public at large in bringing about awareness. Some of these accepted terms with their corresponding meanings are presented in Table 4.2.

## 4.3 Waste Water Reuse Guidelines

A good water reuse system must meet the stipulated guidelines as laid down by the water regulation authorities. The U.S. Environmental Protection Agency (EPA) developed guidelines for water reuse in the year 1992. These include comprehensive information on treating and reusing waste water, and key issues in evaluating waste water reuse opportunities. The main aspects of these guidelines for indirect potable water reuse are stated for three cases.[9]

### 4.3.1 Groundwater Recharged by Spreading into Potable Aquifers

The secondary treatment processes used for this type of IPR include activated sludge processes, trickling filters, rotating biological contactors, and many stabilization pond systems. Secondary treatment should produce effluent in which both the biological oxygen demand (BOD) and suspended solids (SS) do not exceed 30 mg $l^{-1}$. It is also stated that the disinfection may

**Table 4.2** Common terminologies used in waste water reuse.

| S. no. | Terminology | Meaning | Reference |
|---|---|---|---|
| 1. | Fit-for-purpose | The production of reuse water of such a quality that it meets the needs of the end users. | 6 |
| 2. | Waste water | Any water that is derived from a variety of possible uses of the water and usually contains residual pollutants associated with the use of the water. | 7 |
| 3. | Return flows | Treated or untreated waste water that is discharged to a natural surface or groundwater body after use. | 7 |
| 4. | Water reuse | It comprises the use of waste water or effluent from a variety of sources (domestic, industrial, mine effluent) for a new or different beneficial application, such as for drinking, industrial or irrigation purposes. | 7 |
| 5. | Direct Potable Reuse (DPR) | This involves reuse of waste water for drinking purposes after extensive treatment by a number of treatment processes to produce water that is safe for human consumption. | 7 |
| 6. | Non-potable reuse | This involves reuse of treated waste water for purposes other than drinking such as irrigation or industrial use. | 7 |
| 7. | Direct reuse | This involves the use of treated or untreated waste water by direct transfer from the site where it was produced to the site of the new or different beneficial application. | 7 |
| 8. | Indirect Potable Reuse (IPR) | Here, the water is blended with other environmental systems such as a river, reservoir, or groundwater basin, before the water is reused. | 8 |
| 9. | Planned reuse or Intentional reuse | This is the reuse of treated or untreated waste water as a part of the planned project, and it is therefore performed intentionally for a specific application. | 7 |
| 10. | Unplanned reuse or incidental reuse or *de facto* reuse | This is the reuse of treated or untreated waste water after it has been discharged as return flow into a surface water or groundwater body without the intention of reuse, and from which it is then abstracted for a variety of applications. | 7 |
| 11. | Reclaimed water | Reclaimed water (New water) is waste water that has been treated to a level that is suitable for sustainable and safe reuse for the application that it is intended for. | 7 |
| 12. | Recycled water | This is any water that is recovered by treatment of waste water, effluent, grey water or storm water runoff to a quality that makes it suitable for beneficial use. Hence, it is also considered to be a synonym of reclaimed water. | 7 |

be accomplished by chlorination, ozonation, other chemical disinfectants, UV radiation, membrane processes, or other processes. The final water obtained should meet drinking water standards after percolation through the vadose zone.

The EPA mentions that the depth to groundwater (*i.e.* thickness to the vadose zone) should be at least 6 feet (2 m) at the maximum groundwater mounding point and the reclaimed water should be retained underground for at least 6 months prior to extraction. Water that is stored in the aquifers is vulnerable to potential contamination leading to taste and odour issues and high levels of water loss due to evaporation. This is prevented by using soil-aquifer treatment (SAT) and underground storage. During percolation, natural filtration occurs and a drastic improvement in the quality of the infiltrated water occurs. It is necessary that the suggested treatment is site-specific and depends on factors such as type of soil, percolation rate, width of vadose zone, native groundwater quality, and dilution (Figure 4.1).

## 4.3.2 Groundwater Recharged by Injecting into Potable Aquifers

The reclaimed water quality in this case requires the following: pH to be between 6.5 to 8.5, no detectable total coliforms per 100 ml, NTU should be $\leq$2, minimum chlorine residuals should be $\leq$1 mg l$^{-1}$, and total organic carbon (TOC) $\leq$3 mg l$^{-1}$ and total organic halides (TOX) $\leq$0.2 mg l$^{-1}$ are also mandatory. Additionally, the final water should meet drinking water standards. The EPA states that the reclaimed water should be retained underground for at least 9 months prior to withdrawal. It also states that suggested quality limits should be met at the point of injection; monitoring wells are necessary to detect the impact of the recharge operation on the groundwater. A higher chlorine residual content and/or a longer contact time may be needed to assure virus and protozoa inactivation.

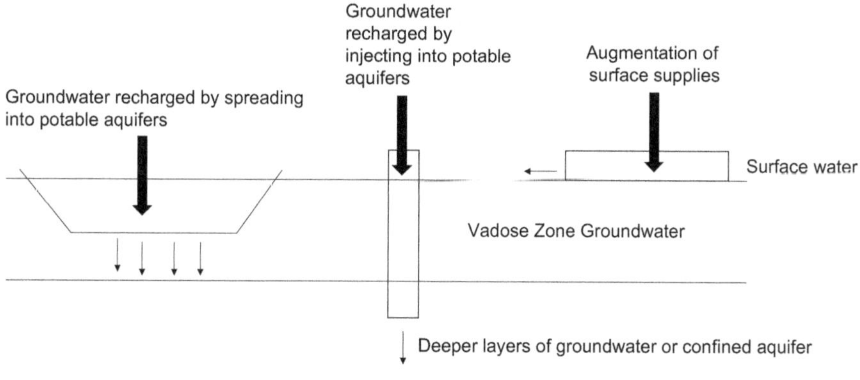


**Figure 4.1** Recharging treated waste water for Indirect Potable Reuse (IPR).

### 4.3.3 Augmentation of Surface Supplies

The reclaimed water quality guidelines for this type of IPR are like those of the groundwater recharged by injecting into potable aquifers. Here, TOC should be $\leq 3 \text{ mg l}^{-1}$ and this type of water should also meet drinking water standards. It is recommended that the level of treatment is site-specific and depends on factors such as receiving water quality, time and distance to point of withdrawal, dilution and following treatment prior to circulation for potable uses. Moreover, the reclaimed water should not contain quantifiable levels of viable pathogens and a higher chlorine ($\leq$1 ppm$\leq$) residual in treated water is recommended to guarantee virus and protozoa inactivation.

## 4.4 Contaminants Generally Found in Waste Water

There is a plethora of contaminants that are generally present in waste water. Mostly, they are microbiological and chemical in nature and depend upon the source of waste water. Although it is beyond the scope of this book to discuss each contaminant in detail, a few of them that are of paramount importance from the water reuse point of view are discussed in this section.

### 4.4.1 Microbiological Contaminants

Microorganisms are ubiquitous in nature and most are not pathogenic to humans. In waste water systems, they act as beneficial decomposers of nutrients and other unstable organic matter. In addition to these beneficial microbes, raw domestic waste water can contain a variety of pathogenic microorganisms that result from feces of infected humans and that are primarily transmitted by the fecal-oral route. Some of the major classes of microorganisms are discussed in this section.

Parasites can be excreted in feces as spores, cysts or eggs (size range from 1 μm to 60 μm), which are resistant to environmental stresses. Helminths can be present as adult organisms, larvae or eggs (size range of 10 to 100 μm). These microorganisms have been found in reclaimed water and are a subject of research owing to the health hazards they pose for humans. *Entamoeba* (amebic dysentery), *Giardia* (Giardiasis or gastroenteritis), *Cyclospora* (cyclosporiasis) and *Toxoplasma* (Toxoplasmosis) are some potentially pathogenic protozoa that need to be inactivated in waste water prior to potable use. Similarly, helminths such as *Ascaris* (Roundworm infection), *Ancylostoma* (hookworm infection), *Taenis* (tapeworm infection) and *Enterobius* (pinworm infection) also require careful monitoring and appropriate treatment.

Yet another important class of microorganisms are bacteria. These are microscopic in size ranging from 0.2 to 10 μm. Several harmless bacteria colonize the human intestinal tract and find their way into waste waters through the feces. Pathogenic bacteria may also be shed along with harmless variants. The number and types of bacteria are a function of their prevalence in the animal and human community from which the waste water is derived. Commonly cited examples are *Shigella* (bacillary dysentery), *Salmonella* (gastroenteritis), *Vibrio cholarea* (Cholera), *Staphylococcus* (skin, eye and ear infection), *Pseudomonas* (infections of the skin, eyes and ear) and *Helicobacter* (ulcers and gastritis).

Viruses, another significant class of microorganisms, occur in various shapes and range in size from 0.01 to 0.3 μm. Enteric viruses multiply in the intestinal tract of human beings and are released in the fecal matter of the infected individual. In general, some viruses persist for only a short period in waste water. Polio, echo, and coxsackie viruses can cause gastroenteritis, heart anomalies, meningitis and respiratory disorders. Hepatitis A and E virus can cause infectious hepatitis, Adenovirus can cause eye infections and respiratory diseases and Rotavirus and Parvovirus can also cause gastroenteritis. These classes of microorganisms can also exist in air in the form of suspended particles. These are called aerosols and their inhalation can cause fatal human infections. Aerosols are often a concern when improperly treated reclaimed water is used for urban and agricultural purposes in the form of sprinkler systems. Sometimes, these aerosols tend to deposit on food, vegetation and other surfaces and result in infections indirectly.[10]

To meet the growing demands for water, it is now imperative to reuse waste water after appropriate treatment. This implies that the level of microbial pollutants present in the waste water must be decreased substantially to render it usable. Needless to emphasize, if the end use is for drinking purposes, then the microbial levels should be nil. Thus, monitoring the microbial levels with cost effective analytical techniques has always been a challenge. Waste water stakeholders have tested the presence or absence of indicator organisms such as *E. coli* and fecal coliforms to predict the possibility of potential pathogen levels since the former are easy and economical to analyze.

However, the disadvantage of using microbial indicators is that they are limited in their ability to predict the presence of pathogens as they require at least 24 h incubation to grow in conventional media. Moreover, coliform bacteria being highly susceptible to chemical disinfection fail to correlate with protozoan parasites such as *Cryptosporidium* and enteric viruses that may be present in disinfected reclaimed waste water. Thus, alternative microbial indicators have been suggested for their evaluation in waste water but there are only a few studies reported and additional study in this area is needed.[11]

Yet another aspect is the evolution of better analytical methods for actual pathogen monitoring and there have been several studies carried out that have not relied solely on conventional culture methods in microbiological media.[12] New molecular biological methods are under development that enable a quicker determination of specific microorganisms. One such example is Polymerase Chain Reaction (PCR), which is now commonly used to study pathogens and indicators by detecting the DNA or RNA in the water samples. It is a very sensitive method and requires less time as compared to conventional techniques of growing microbes in culture media. Many studies report the use of PCR in the analyses and determination of pathogens in recycled waste water.[13] On the flip side, PCR techniques only indicate the presence or absence of DNA or RNA derived from microorganisms and they do not determine pathogen viability or infectivity. Quantitative PCR can provide data for quantitative microbial risk assessment (QMRA). However, this process may not be cost effective as compared to the conventional culture techniques used to detect microbial populations in each water sample.[10]

### 4.4.2 Chemical Contaminants

Any type of water usually contains some chemical constituents, which, if present in more than certain concentrations, can prove harmful to humans and crops. For instance, rain water collects chemicals from the atmosphere, groundwater contains inorganic substances from earth and surface water also contains many natural substances, pesticides and chemicals from industrial discharges. Similarly, waste water contains myriad chemical substances and the amount of such substances depends on factors such as the source of the waste water, collection system and the treatment processes used.[10] Chemicals in low concentrations may not have any direct toxic effects. However, over a period, they can lead to bioaccumulation and long term chronic problems.

Therefore, it is imperative that the water regulatory bodies have included a list of chemical parameters to regulate their levels in the reuse of reclaimed waste water. An extensive list of micropollutants has been discussed since it is known that a major number of trace chemical pollutants persist in municipal waste water even after conventional treatment processes. Several pollutants such as the endocrine disrupting chemicals, pesticides and polar pharmaceuticals are often present in waste water and require proper monitoring and treatment to prevent potential health risks to humans and damage to crops depending on its end use. Some classes of chemical contaminants normally encountered during reuse of waste water are discussed in this section.

Inorganic chemical constituents in waste water include metals, salts, oxyhalides, nutrients and engineered nanomaterials. The aggregate measures of most of these constituents in water are denoted as Total Dissolved Solids (TDS) and conductivity. Boron, a metalloid in detergents can be toxic to plants at concentrations that may be present in reclaimed water. This can limit its use for irrigation of plants. Hence, a limit of 0.4 to 1.0 $mg\,L^{-1}$ Boron in treated waste water for irrigation purposes and aquaculture and 0.2 $mg\,L^{-1}$ for indirect aquifer recharge has been proposed in the guidelines from the EPA. Oxyhalides of alarm in water reuse are bromate, chlorate and perchlorate that must be treated depending on the end use. Nitrogen and Phosphorus from human waste can cause environmental and health issues but can also be beneficial in irrigation applications. Therefore, treatment of such waste waters will depend on the intended use of the product water.

Nanomaterials are yet another class of contaminants that can be present in waste water given the fact that many nanomaterials such as nanomembranes are used for water treatment. However, there are no reports on any adverse human health effects. The potential ecotoxicological risk posed by the release of nanoparticles to surface water highlights the need for its restriction in the use of nanomaterial-containing commercial products. Although limited research has been carried out on their fate in waste water treatment, preliminary findings indicate that engineered nanoparticles will associate with bio solids or remain in effluents,[14] leading to adverse long term health effects.

Organic chemical contaminants in waste water include humic substances, fecal matter, kitchen waste, oils, liquid detergents and substances that eventually become a part of the sewage system. These can cause colour and odour problems, and result in the clogging of pipes and sprinkler heads. Organics are also a good source of nutrients for microorganisms and thus lead to an increase in the microbial population warranting further treatment. On the other hand, organics can interfere with chemical disinfectants such as chlorine and ozone and UV disinfection, rendering them less effective. Moreover, chlorination of water containing organic substances can result in the formation of chlorinated disinfection byproducts (DBPs), which are potentially harmful to human health and can result in diseases such as bladder cancer.[15] Similarly, chloroform, commonly found Trihalomethane (THM) compounds and Haloacteic acids (HAAs) are also undesirable byproducts of chlorine disinfection and are carcinogens. The carcinogen *N*-nitroso dimethylamine (NDMA) is also produced when reclaimed water is disinfected with chlorine or chloramines.[16]

## 4.5 Treatment Methods

Since this book is about drinking water treatment for developing countries, this section will essentially describe various treatment technologies for DPR and IPR. It is interesting to note that historically, the goal of waste water treatment has always been meeting the water quality limits as per the guidelines laid down by water authorities, which was related to the discharge permit requirements. For potable reuse projects, the waste water treatment plant becomes a part of the overall treatment scheme and therefore treatment goals should focus on providing a water quality that will assist the downstream advanced treatment processes. Flow equalization, improved primary and secondary treatment, increased solid retention times and alternate management of return flows from solid processing are some of the important modifications that change the pattern of waste water treatment to benefit potable reuse treatment trains.

### 4.5.1 Indirect Potable Reuse

IPR is a very important water reuse option where municipal waste water is extensively treated and discharged into the groundwater or surface water sources to augment the potable water supply. IPR is considered as a feasible alternative in many countries globally as it is not dependent on natural factors such as rainfall. Additionally, a very high quality of water suitable for human consumption can be generated. However, the main challenge is the health risk involved and therefore it warrants a cautious approach. Hence, careful evaluation of possible human and environmental risks as well as the active participation of the community are mandatory factors in IPR implementation. This section essentially discusses the planned IPR where the treated waste water is intentionally added to the water supply system. Some of the IPR projects from around the globe are described as follows (Figure 4.2).[17]

#### *4.5.1.1 Orange County Water District (OCWD) Project, California, USA (1975–2004)*

This project was built in the year 1975 and decommissioned in 2004 to be upgraded later as the ground water replenishment system (GRS) system described here. This was the first project that employed recycled water to maintain a sea water intrusion barrier and more than half the injected water increases the potable water supplies. The treatment technology included lime clarification, recarbonation, multimedia filtration, granular activated carbon, filtration and chlorination. Reverse osmosis was added in the year 1977 and advanced oxidation with hydrogen peroxide and UV was included around 2001. The OCWD ground water replenishment system (GRS), which was an upgrade of the existing project, was built from 2004 to 2007 and it has been working as a full-scale plant since then. Microfiltration and Reverse Osmosis and advanced oxidation systems that include UV and hydrogen peroxide were included for treatment. The full-scale plant produces around 10% of Orange County's potable water. Gradually, 100% recycled water was injected over time and the groundwater aquifer now supplies more than half of the population.

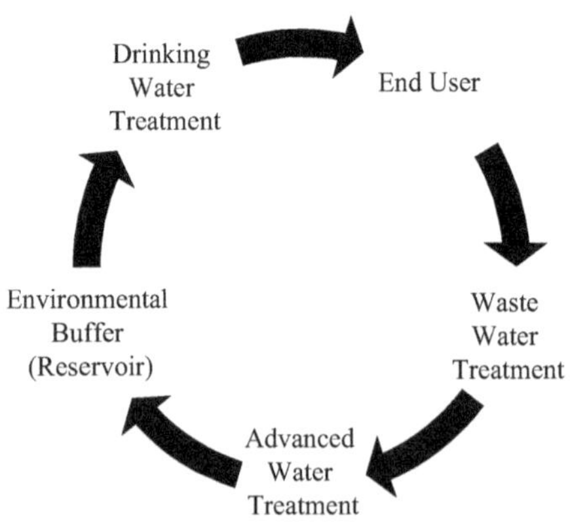


**Figure 4.2** Indirect potable reuse.

#### *4.5.1.2 Denver Potable Water Demonstration Project, Colorado, USA (1985–1992)*

This project studied different possibilities for water reuse and found that potable reuse was a viable option. The pilot plant used unchlorinated secondary effluent from the Denver Waste Water treatment plant. Various treatment technologies used were: high-pH lime clarification, sedimentation, recarbonation, filtration, selective ion exchange for ammonia removal, UV irradiation, activated carbon adsorption, RO, air stripping, ozonation, chlorine dioxide disinfection, ultrafiltration and chloramination (Figure 4.3).

#### *4.5.1.3 San Diego Water Repurification Project, California, USA (Since 1981)*

This project was set up in 1981, and eventually in 1995, several treatment methods were tested including reverse osmosis and granular activate carbon filter. Since the year 2000, microfiltration and reverse osmosis and advanced oxidation using UV light and hydrogen peroxide have been used. The first demonstration plant was established between 1985 and 1999, and since 2000, the full scale project was set up for non-potable reuse only. This was mainly due to non-acceptance by the community.

#### *4.5.1.4 The Chelmer Augmentation of Wastewater Reuse Scheme, Essex, England (1997)*

In this project, recycled water was discharged into the Chelmer river, which is used to augment the Hanningfield reservoir, with 8–12% blending with treated waste water. The reservoir storage time was typically around 200 days. The main treatment used was microfiltration and UV radiation.

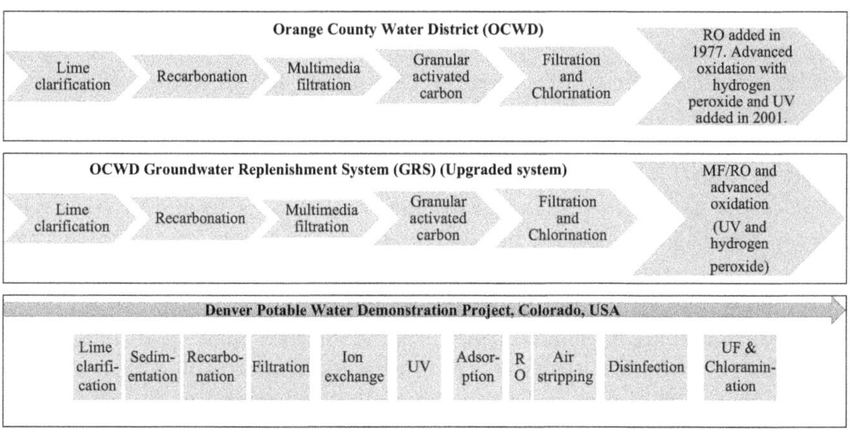


**Figure 4.3** Examples of IPR projects – I.

The treated water has been monitored for viruses and estrogens since the year 1996. This project caters to a population of around 1.7 million.

#### 4.5.1.5 *Water Reclamation and Study (NeWater), Singapore (2000)*

This was initially operated as a demonstration plant, and since 2000, it has been operated as a full-scale plant. This happened when adoption of augmentation of the potable water supply was recommended. Now, it has three operating plants with a total production capacity of 92 ML per day and most of the recycled water is used for industrial purposes. The main highlight of this project was the use of a well-documented community education programme. The treatment methods used are ultrafiltration, reverse osmosis, UV and chlorination.

#### 4.5.1.6 *Torrie Reuse Plant, Wulpen, Belgium (2002)*

This plant mainly employs Microfiltration and Reverse Osmosis and UV disinfection to produce recycled water that is blended up to 40% with the drinking water supplies. It is thus a full-scale project that serves around 60 000 people and meets between 40 to 50% of their drinking water demands. The treated water is stored in a reservoir for a period of 40 days. The drinking water quality obtained is reported to have a lower hardness and better colour due to the reduced organic content (Figure 4.4).

Literature reports have described the varied use of treatment technologies for full scale IPR schemes that range from soil application of raw waste water to highly engineered membrane systems combined with advance oxidation processes. From the preceding examples of IPR projects, in any IPR project, a general treatment scheme of ultrafiltration or microfiltration is usually carried out prior to reverse osmosis, which is subsequently followed by UV treatment or an advanced oxidation process. In the microfiltration step, the effluent from the waste water treatment plants passes through a membrane with a pore size of 0.01 μm, which ensures the removal of most of the suspended solids, bacteria and protozoa. During the RO step where specialized membranes are employed, rejection of most of the high molecular weight organic matter occurs. High rates of BOD and chemical oxygen demand (COD) removal also happen during this stage and reverse osmosis (RO) achieves the separation of minerals, heavy metals, viruses and even pesticides, rendering the waste water fit for IPR.

It is not just the treatment process but also its continuous monitoring that makes any IPR project a success in the long run. Since human health must be protected, it also becomes important that each treatment step is evaluated for different types of contaminants to perform a timely check. For instance, additional treatment barriers after the reverse osmosis step were introduced in the GRS (OCWD) project when potential carcinogens such as NDMA and 1,4 dioxane were detected. Therefore, monitoring the

**Figure 4.4** Examples of IPR projects – II.

performance of each treatment barrier in any multi barrier treatment approach can ensure safe drinking water.

It is interesting to note that there has been ongoing research to improve the treatment schemes in IPR and one of the emerging technologies is nanofiltration. With a greater ability to remove most of the microbial and chemical pollutants, nanofiltration is touted as a substitute for reverse osmosis as it operates at much lower pressures, reducing the operating cost, and gives significantly higher flux. It can also remove trace contaminants, which usually are a major cause of concern. Comparative studies with reverse osmosis have revealed that medium to high removal rates have been obtained for endocrine disrupting compounds, pesticides and certain pharmaceutical compounds.

## 4.5.2 Direct Potable Reuse

In contrast to IPR, in the case of direct potable reuse, there is no requirement of release to groundwater or surface water reservoirs, which are often termed as environmental buffer systems.

Since the blending of treated water with other drinking water sources is less costly than in the case of IPR, DPR is usually preferred in many scenarios for the ease of operation and economics. Therefore, over the last few years, a lot of research has been undertaken worldwide in the field of DPR. Concurrently, areas such as hazard identification, validation of control measures, case studies, public acceptance and legal aspects in the case of DPR are also receiving a lot of attention. Some of the DPR projects are listed in Table 4.3.

The requirements of any potential DPR treatment scheme are manifold and the specific treatment method selected should take into consideration the specific water quality characteristics, experience of the water system

**Table 4.3** Examples of DPR projects.

| S. no. | DPR project name and location | Source water | Treatment process | Guidelines followed | Major pollutants – levels before and after treatment | Reference |
|---|---|---|---|---|---|---|
| 1. | New Goreangab Water Reclamation Plant (NGWRP), in Windhoek, Namibia (2002) | Treated effluent from municipal waste water treatment plant | Powdered activated carbon (PAC) dosing, pre-oxidation and pre-ozonation, flash mixing, enhanced coagulation and flocculation, dissolved air flotation, dual media rapid gravity sand filtration, ozonation, BAC filtration, GAC filtration, ultra-filtration (UF), disinfection and stabilisation | WHO Guidelines (1993), Rand Water (South Africa) Potable Water Quality Criteria (1996) and the Namibian Guidelines for Group A water (NamWater, 1998) | Turbidity (NTU) from 53 to 0.1 *Giardia* per 100 mL from 214 to 0 *Cryptosporidium* per 100 mL from 334 to 0 *E. coli* per 100 mL from 20 347 to 0 Heterotrophic Plate Count per 1 mL from 332 150 to 808 Iron from 2.8 to 0.05 mg $L^{-1}$ | 18 |
| 2. | Brownwood, Texas, USA Cl2, UF, UV, NH3, dechlorination, RO, GAC, UV, NH3, Cl2 | Tertiary municipal effluent | Chlorination, UF, UV-disinfection (stage 1), chloramination, de-chlorination, RO, activated carbon filtration, UV disinfection (stage 2) and chloramination | Texas Commission on Environmental Quality (TCEQ) | Final values TDS – 1000 mg $L^{-1}$ Fluorides – 2 mg $L^{-1}$ | 19 |
| 3. | Beaufort West, South Africa (2011) | Treated effluent from municipal waste water treatment plant | Phosphate removal with ferric chloride, sedimentation, pre disinfection with chlorine, sand filtration, ultrafiltration, Reverse osmosis, advanced oxidation with $H_2O_2$ and UV, pH stabilization and disinfection with chlorine | South African National Standard (SANS) 241–1 : 2011 | Final Values after treatment: Turbidity (NTU) −0.2 Fluoride (mg $L^{-1}$) <0.1 Fecal coliforms (count per 100 ml) – Not detected | 20 |

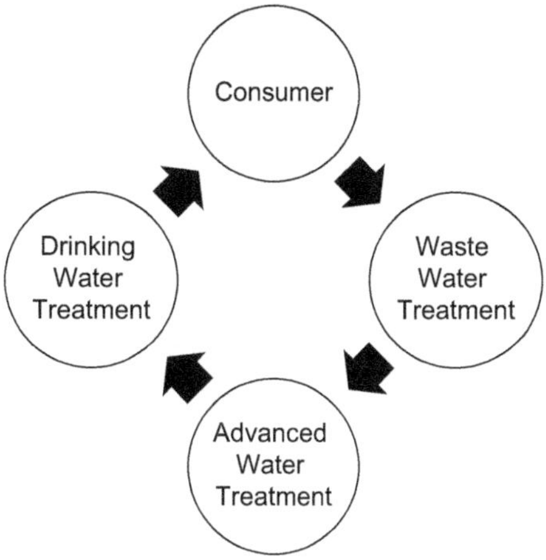


**Figure 4.5** Direct potable reuse.

operators, extent of fool proof systems and protocols and the amount and frequency of monitoring. The default treatment schemes adopted for direct potable reuse appear to be similar globally and are shown in Figure 4.5. Most of these plants are membrane based systems that employ microfiltration or ultrafiltration followed by reverse osmosis, which in turn is followed by treatment with UV or an advanced oxidation process.

Increasing numbers of research studies indicate that although this treatment scheme offers a very high quality of water that is safe for human consumption, the RO systems considerably increase the cost of the treatment schemes. The main reason is the huge amount of energy consumption by typical RO processes and the management of the concentrate produced. Therefore, there appears to be a great interest in the use of DPR systems that do not use the RO step. Alternatively, as with the IPR systems discussed in the previous section, here too there is an increased interest in the use of nanofiltration instead of RO to reduce the energy consumption. Treatment schemes that do not use RO usually employ an activated carbon adsorption stage along with an ozonation step. Very often, such systems have been found to be effective in removing trace organic compounds in water to below detection levels.

It is interesting to note that several studies that compared the two treatment schemes, *viz.* MF-RO-UV and Ozone-BAC-GAC-UV that are commonly used in DPR, indicate that the capital and operating costs in the case of the former (RO based) were almost 50–60% higher than those of the latter (non RO based). Moreover, the energy costs while using the RO based systems were appreciably higher, thus making the non-RO based systems much more sustainable. Further studies on membrane systems have led to the concept of using forward osmosis (FO), a promising membrane technology. The idea here is to combine seawater desalination and water reuse and this hybrid is called the FO–RO system. In this process, high quality water recovered from a waste water stream is used to dilute the seawater before reverse osmosis treatment.

The advantages include lower desalination energy requirements, water augmentation and delivering safe water for DPR. Although the hybrid systems have drawbacks such as public acceptance, water management challenges and technical issues such as low FO permeation flux, recent developments in the form of improved membrane systems for FO have made it simpler to adopt such methods. Although there is a need for a detailed study, especially on a full-scale FO–RO scheme and the economics involved, such a hybrid system is indeed very promising.[21]

## 4.6 Economic Considerations

The costing of potable water reuse systems varies across different regions and depends on whether they are IPR or DPR and the type of treatment trains employed. For instance, the Australian Academy of Technological Science and Engineering (ATSE) reported that a hypothetical DPR system would cost US$616 million in terms of capital costs as compared to US$1287 million for an IPR system having a treatment capacity of 120 million litres per day (MLD). The operating costs were reported to be US$53 million per year for DPR and US$72 million per year for IPR systems.[22] This clearly points out that DPR systems are cheaper as compared to IPR.

According to the Texas Water Development board, different treatment schemes have different costs owing to the different unit operations involved. In one treatment scheme, which employs Secondary/Tertiary, Ozone, MF/UF, RO, UV/AOP and Stabilization, the capital cost for treating 25 million gallons per day (MGD) is US$5.13 M/MGD and the operating and maintenance cost is US$0.610 M/MGD. Another treatment scheme that uses a different method, *i.e.* Secondary/Tertiary, Ozone, BAC, UV, Engineered Storage and water Treatment Process (WTP), had capital and operating and maintenance costs for the same capacity of US$1.42 M/MGD and US$0.103 M/MGD, respectively.[23] Thus, we can assess the difference in costs based on the methods used.

## 4.7 Merits and Demerits

Both IPR and DPR have several pros and cons and adoption of any one process will depend on weighing these before it is successfully implemented and used by the end consumer. In the case of IPR, the advantages include environmental purification as in the case of IPR, the treated water is discharged into a water body such as a river or lake and this allows any remaining pollutants to be degraded by natural processes. IPR appears to be well established globally, and presently, its public acceptance is also higher as compared to DPR. On the other hand, IPR systems are more expensive and rely on enclosed pipelines and pumping systems over long distances. Many experts feel that IPR processes are inefficient as they treat the same water twice (waste water treatment also includes RO and advanced oxidation, which is again carried out during IPR). The perceived health issues and the perception that the quality of a natural water body is being degraded by the addition of reclaimed water still appear to affect public acceptance to a great degree.

The main merits of DPR appear to be the cost savings as drinking water treatment and waste water treatment typically occur in nearby locations, thereby DPR systems require less pumping of water to end consumers. This in turn reduces the energy requirements and the greenhouse gas emissions. Additionally, owing to the short distance that water must travel, the DPR systems are less affected by damages due to floods or human made disasters. The disadvantages of DPR systems are safety concerns over their ability to fully remove certain compounds such as pharmaceutical residues, personal care products and certain endocrine disrupting compounds from water. Some DPR systems may require additional set-up, thereby increasing the cost of the treatment. Last but not least, public acceptance is still a major hurdle in the implementation of DPR projects and a lot of public education and outreach programs are needed to make consumers feel positive about DPR.[24]

## 4.8 Conclusions

The major water crunch that is being faced by mankind is driving us to explore new sources for safe drinking water and waste water reuse is one of the major means of achieving that. Both IPR and DPR are very viable options for supplying safe drinking water to those regions with increased water demands. No water treatment is ever without risk, including conventional water treatment. Similarly, IPR and DPR will also never be totally risk free. However, with the best technologies, risk assessment and management practices, the potential chemical and microbial contaminants in water can be inactivated and safe potable water can be made available. IPR will represent an important part of sustainable water resource management in many regions globally. DPR is becoming more economical as compared to IPR, as discussed earlier in this chapter, and in general, it can be stated that DPR could become a commonly practiced water management option in the years to come.

## 4.9 Further Research and Recommendations

1. There is a lot of scope for improvement in the existing treatment technologies that are used in IPR and DPR. Future research should investigate ways to enhance these methods and explore novel techniques for safe potable reuse that are cost effective and simple to implement, especially in developing countries.
2. With the advent of newer and emerging contaminants in water systems, detection of these pollutants will play an important role in the selection of an appropriate treatment train, both in IPR and DPR as well. Developing better and innovative monitoring methods to help detect potential contaminants that can pose health risks is the need of the hour.

3. Public education and convincing more and more people to adopt potable reuse systems will be the major challenge in times to come. Innovative and creative ways to reach out to the population will need a lot of focus by various stakeholders worldwide.
4. Appropriate guidelines addressing major contaminants and their acceptable limits with respect to DPR need to be drafted meticulously by the water regulating authorities on a global as well as local basis.

# References


1. http://www.nesc.wvu.edu/ndwc/articles/OT/WI05/reuse.pdf (last accessed February 2017).
2. https://www3.epa.gov/region9/water/recycling/ (last accessed February 2017).
3. L. Onyango, F. Leusch, G. Leslie and J. G. Wood, *Water Quality and Public Health: Risks and Prevention, Health Assessments and Potable Use Case Studies*, Australian Water Recycling Centre of Excellence, Brisbane, Australia, 2015.
4. *Water Reuse: An International Survey of Current Practice, Issues and Needs, Scientific and Technical Report No. 20*, ed. B. Jimenez and T. Asano, IWA Publishing, London, UK, 2008, [Status of water reuse].
5. http://unesdoc.unesco.org/images/0024/002471/247153e.pdf.
6. V. Lazarova, T. Asano, A. Bahri and J. Anderson, *Milestones in Water Reuse. The Best Success Stories*, IWA Publishing, London, UK, 2013.
7. http://greencape.co.za/assets/Sector-files/water/Reclamation-and-reuse/WRC-Decision-support-model-for-the-selection-and-costing-of-direct-potable-reuse-systems-from-municipal-wastewater-2014.pdf (last accessed January 2017).
8. https://watereuse.org/water-reuse-101/glossary/ (last accessed February 2017).
9. http://www.lacsd.org/civicax/filebank/blobdload.aspx?blobid=2184 (last accessed April 2017).
10. https://nepis.epa.gov/Adobe/PDF/P100FS7K.pdf (last accessed April 2017).
11. V. J. Harwood, A. D. Levine, T. M. Scott, V. Chivukula, J. Lukasik, S. R. Farrah and J. B. Rose, *Appl. Environ. Microbiol.*, 2005, **71**(6), 3163.
12. P. Fox and J. E. Drewes, *J. Environ. Assessment Monit.*, 2001, **70**, 117.
13. P. K. Jjemba, L. Weinrich, W. Cheng, E. Giraldo and M. W. LeChevallier, *Guidance Document on the Microbiological Quality and Biostability of Reclaimed Water and Distribution Systems (WRF-05-002)*, WateReuse Research Foundation, Alexandria, VA, 2010.
14. R. Kaegi, A. Voegelin, B. Sinnet, S. Zuleeg, H. Hagendorfer, M. Burkhardt and H. Siegris, *Environ. Sci. Technol.*, 2011, **45**(9), 3902.
15. K. S. Freeman, *Environ. Health Perspect.*, 2010, **118**(11), 118.
16. W. A. Mitch, A. C. Gerecke and D. L. Sedlak, *Water Res.*, 2003, **37**, 3733.

17. C. Rodriguez, P. V. Buynder, R. Lugg, P. Blair, B. Devine, A. Cook and P. Weinstein, *Int. J. Environ. Res. Public Health*, 2009, **6**, 1174–1209.
18. J. Lahnsteiner and G. Lempert, *Water Sci. Technol.*, 2007, **55**, 441–448.
19. J. Lahnsteiner, P. L. du Pisani, J. Menge and J. Esterhuizen, More than 40 years of direct potable reuse experience in Windhoek, Namibia, in *Milestones in Water Reuse. The Best Success Stories*, ed. V. Lazarova, T. Asano, A. Bahri and J. Anderson, IWA Publishing, London, UK, 2013, pp. 351–364.
20. http://www.imesa.org.za/wp-content/uploads/2015/08/Paper-6.pdf (last accessed April 2017).
21. G. Blandin, A. R. D. Verliefde, J. Comas, I. Rodriguez-Roda and P. Le-Clech, *Membranes*, 2016, **6**, 37.
22. http://www.australianwaterrecycling.com.au/_literature_128048/Drinking_Water_Through_Recycling_-_Appendix_B (last accessed April 2017).
23. http://www.twdb.texas.gov/publications/reports/contracted_reports/doc/1248321508_Vol1.pdf (last accessed April 2017).
24. https://www.wateronline.com/doc/direct-potable-reuse-vs-indirect-weighing-the-pros-and-cons-0001 (last accessed March 2017).

CHAPTER 5

# *Case Studies*

Drinking water treatment in developing countries is very essential to prevent the risk of consuming contaminated water from sources that are not well maintained or likely to be polluted from effluents or sewage. This has been much elaborated in the introductory chapters and therefore a spectrum of water treatment techniques particularly suited for developing nations have been dealt with in detail in the previous chapters. Not only the treatment of prevailing water sources but conservation of water through rain water harvesting and its treatment has also been amply highlighted. The possibility of waste water reuse by treating it to make it suitable for direct or indirect potable uses and its customization for developing countries have also been emphasized in the preceding chapters.

To understand the above aspects in a better way, it is essential to cite specific examples or cases where a pollutant has been successfully treated and removed and the water rendered safe for human consumption. Since it will be beyond the scope of this book to deal with every pollutant, the aim of this chapter is to discuss a few instances where either the pollutants are efficaciously removed from an unclean water body or one or more water conservation practices have made clean and safe water easily accessible to the people in the developing nations. Thus, six case studies are presented here.

These six case studies are (i) Arsenic: A Looming Danger. Can we do something? (ii) Drumstick (*Moringa oleifera*) for water purification; (iii) Rain water harvesting in a developing country; (iv) Waste Water Reuse for direct/indirect potable use in a developing country; (v) Lake Water Treatment in Developing Countries; (vi) Disinfection of borewell water using hydrodynamic cavitation in India Mark II Hand Pump, which are expected to be typically representative. These case studies have been selected based on their relevance and because they have been discussed in detail in the previous

Drinking Water Treatment for Developing Countries: Physical, Chemical and Biological Pollutants
By Aniruddha Bhalchandra Pandit and Jyoti Kishen Kumar

Published by the Royal Society of Chemistry, www.rsc.org

chapters. Therefore, citing an example with quantification under each of these topics will be useful and meaningful in the context of drinking water treatment in developing countries.

# 5.1 Case Study 1 – Arsenic: A Looming Danger. Can We Do Something?

## 5.1.1 Arsenic in Drinking Water and the Magnitude of the Problem

Arsenic is a chemical element that occurs either as a pure element or in many minerals generally in combination with other metals and sulfur. It is a metalloid, *i.e.* it has properties inbetween those of metals and nonmetals, and it is widely distributed throughout Earth's crust. In water, it is most likely to be present as arsenate with an oxidation state of 5 if the water is oxygenated. It may be also found under reducing conditions as Arsenic(III) and such conditions are generally encountered in deep lake sediments or in groundwater systems. It is interesting to note that an increase in pH is likely to increase the concentration of dissolved arsenic in water.

Among its various allotropes, the gray form is regarded as important for industrial applications. Arsenic is used as an alloying agent in the processing of glass, textiles, paper, wood preservatives and ammunition. It is also used in the production of lasers, transistors and semiconductors. Arsenic also finds application in the hide tanning process and to a limited extent in pesticides.

There are several routes by which arsenic may get introduced into water. It could be from industrial effluents such as mining wastes, through dissolution of rocks, minerals and ores and *via* atmospheric deposition. Based on the source of water, the arsenic concentrations may vary. For instance, the arsenic levels in natural waters like the sea and oceans is usually around 1 to 2 $\mu g\, l^{-1}$. In regions where there are natural sources of arsenic, the levels may increase to as high as 12 $mg\, l^{-1}$ such as water near mining and agrochemical manufacturing facilities. It is alarming to know that the level of arsenic in geothermal waters can be as high as 25 $mg\, l^{-1}$ and elevated levels of arsenic in drinking water can lead to a plethora of health risks.

Arsenic is reported to be a carcinogen by the WHO, and The International Agency for Research on Cancer (IARC) has also classified arsenic in drinking water to be carcinogenic to humans. It is one of the most significant chemical pollutants found in potable water worldwide. The WHO has now recognized that almost 140 million people in 50 countries have been drinking water having arsenic at elevated levels, *i.e.* much above the provisional guideline value of 10 $\mu g\, L^{-1}$. For instance, in the year 2012, it was reported that around 39 million people in Bangladesh, India were still exposed to arsenic concentrations above the national standard of 50 $\mu g\, L^{-1}$ and the WHO guideline value of 10 $\mu g\, L^{-1}$.

### 5.1.2 Health Risks

Ingestion of arsenic containing drinking water may lead to acute symptoms such as vomiting, nausea, diarrhea and abdominal pain, which is also followed by muscle cramps, numbness of extremities and, in extreme cases, it can be fatal. On the other hand, long term exposure (more than 5 years) caused by the consumption of arsenic laden drinking water for a prolonged period may lead to dermatological issues. These include skin pigmentation changes, lesions and hard patches on the palms and feet, which in turn can lead to skin cancer. Apart from this, chronic arsenic exposure can also lead to lung and bladder cancer.

Other major long term health issues include the development of other health problems such as diabetes, cardiovascular issues and pulmonary diseases. It is distressing to note that arsenic in drinking water is also linked with ‘Blackfoot disease’, which is a severe condition of blood vessels leading to gangrene. This has been reported in the province of Taiwan and is often also linked with malnutrition among the population, which aggravates this condition. Among others, arsenic toxicity can lead to adverse pregnancy outcomes and cause infant mortality and child death.

Several studies carried out across the globe have revealed that drinking water with arsenic above 50 $mg\,l^{-1}$ during pregnancy leads to augmented risks of fetal loss and infant death. An important revelation from these investigations is that the arsenic exposed group (who were pregnant) with severe skin lesions had higher risks of developing adverse effects in their pregnancy. Arsenic toxicity appears to alarmingly affect children and many studies have showed that drinking water with elevated levels of arsenic affected the intelligence quotient and functioning among children below 11 years of age and these children also had a higher risk of developing cancer. However, since there are no reliable methods to distinguish the cancer caused by arsenic from the cancers caused by other factors, there appears to be no dependable estimate of the extent of this problem globally. The US national Research Council reports that 1 in 100 additional cancer deaths could be expected from a lifetime exposure to arsenic in drinking water with levels above 50 $\mu g\,L^{-1}$. This clearly points out the grave situation and thus, immediate action is required to combat Arsenic poisoning in drinking water to save the affected population.

### 5.1.3 Treatment Methods

Arsenic remediation methods are highly researched worldwide and a plethora of techniques are available today. The focus of this case study is to elaborate arsenic removal techniques in terms of their efficiency of Arsenic removal, cost effectiveness, scale of operation and feasibility in developing countries. But before that, it is important to become familiarized with the conventional arsenic removal technologies that have been employed for a long time. Some of the arsenic removal methods include oxidation/precipitation, coagulation, sorption, ion exchange, membrane techniques and hybrid methods using liquid filters with hand pumps.[1] Fascinatingly, novel methods for arsenic removal keep emerging from various parts of the world. The following section briefly highlights these technologies.

### 5.1.3.1 *Conventional Methods of Arsenic Removal*

It is important to note that all the methods for arsenic removal are usually based on some chemical processes such as oxidation or reduction reactions that alter the chemical form of arsenic. Thus, oxidation involves the conversion of soluble arsenite to arsenate, which is then removed by coagulation and precipitation processes or adsorption and/or ion exchange. Thus, oxidation does not remove the arsenic from water, but it aids in the optimization of the overall process by converting arsenic to an insoluble form. Although there is a plethora of oxidants that have been used for oxidation in arsenic removal, atmospheric oxygen, hypochlorite and permanganate are most commonly used in developing countries as they are relatively low-cost chemicals and are easily available. Generally, oxidation of arsenite with oxygen is reported to be a very slow process and therefore chemicals such as chlorine, ozone or permanganate are adopted. The choice of the chemicals used is largely dependent on the nature of other substances present in water. For instance, the presence of organic material in water was found to decrease the oxidation rates of arsenite. Therefore, the oxidants need to be selected carefully and as discussed earlier, oxidation as a sole technology for arsenic removal is not beneficial.

Coagulation and flocculation are another arsenic removal technique that is frequently used that causes the dissolved arsenic to form a low solubility solid like calcium arsenate. This is then easily removed by sedimentation and filtration. This is one of the most documented methods for arsenic removal. In the case of coagulation, coagulants are introduced into water containing arsenic. Generally, coagulants such as ferric chloride or aluminum sulphate that bear positive charges are used, which reduce the negative charges of the colloids and make the particles clump together. Flocculation is the addition of anionic flocculants that result in charge neutralization between particles. Subsequently, larger flocs are formed, and these can be easily removed by techniques such as filtration or sedimentation. It is generally reported that arsenate is more efficiently removed as compared to the arsenite form of arsenic in water.

Yet another vital process for arsenic removal is adsorption and ion exchange. In these processes, a solid adsorbent is used that successfully adsorbs the arsenic from water by forces such as electrostatic forces and van der Waals forces between the adsorbent and arsenic molecules. Several adsorbents have been used such as activated carbon, coal, zeolites, activated alumina, iron hydroxide, titanium dioxide and cation exchange resins. Although each of these adsorbents has different properties, they all need to have a high surface area and porosity for adsorption of arsenic from water. Recently, iron-based adsorbents have received great attention by experts globally due to the ability of iron to remove arsenic either by adsorption or co-precipitation. This high affinity between iron hydroxides and arsenic has been exploited successfully for arsenic removal in different parts of the world and one such case study is presented in the succeeding section. More information on this technology with its pros and cons and cost effectiveness is highlighted in the case study.

Membrane filtration is a well-known and effective method for removal of pollutants from water and it has been found useful in the removal of arsenic from water. Membranes are prepared synthetically with specific pore sizes that act as barriers and selectively allow certain molecules to flow through and retain certain contaminants on the membrane filter surface. Selection of a specific membrane with an apt pore size to suit its application is an important factor to be emphasized. Membrane processes that require low operating or transmembrane pressures such as microfiltration and ultrafiltration and high-pressure filtration methods such as reverse osmosis and nanofiltration can be used to remove arsenic from water. However, one important aspect is the need to chemically convert arsenic into a form that can be easily filtered by the membrane processes.

Most frequently, oxidation of arsenite (it has a neutral charge in the pH range of 4 to 10) to arsenate (it has a negative charge in the pH range of 4 to 10 and can easily bind onto flocculant surfaces) and flocculation using ferric chloride or ferric sulphate are carried out before microfiltration. Similarly, pre-treatment of arsenic containing water is required before subjecting it to ultrafiltration. Generally, treatment with cationic surfactants such as hexadecylpyridinium chloride (CPC) is carried out that results in the formation of micelles, which can bind to negatively charged arsenic species. These surfactant aggregates can then be easily removed by ultrafiltration. However, a major drawback here is the high levels of surfactants that will be present in the permeate, necessitating further treatment before the end use of water. Nanofiltration and reverse osmosis also significantly reduce arsenic levels in water, provided water is free from other suspended pollutants and arsenic should be in the form of arsenate.

### *5.1.3.2 Emerging Technologies*

Apart from the traditional methods for arsenic removal, several newer technologies that are based on novel concepts or that enhance the existing techniques have been investigated by experts globally. This is mainly due to the drawbacks and limitations of the conventional methods, as discussed previously, and with an aim to provide relief to populations affected with arsenic toxicity by way of simple and cost-effective technologies. The literature has ample examples of newer techniques for arsenic removal and some of the recent ones are highlighted in Table 5.1. It is important to understand that many such emerging methods have only been tried on a lab scale or are

**Table 5.1** Some emerging technologies for arsenic removal from drinking water.

| S. no. | Technology | Salient features | Reference |
|---|---|---|---|
| 1. | Nanoparticles | Adsorption is the main mechanism. Small size, large surface area and high adsorption capacity as compared to conventional adsorbents make nanomaterials a promising choice. Moreover, they can be functionalized for specific requirements. There have been studies at bench scale only and its cost effectiveness and waste disposal issues must be studied further before it can be scaled up. | 1 |
| 2. | Water treatment residuals (WTRs) as adsorbents | These are essentially byproducts of de-ironization and de-manganization processes of infiltration water. They mainly consist of iron and manganese oxides usually in the ratio of 5 : 1 and have a high surface area that adsorbs arsenic. | 13 |
| 3. | Aluminium (Al) based sorbents | Aluminium based coagulants are a good choice for arsenic removal due to their low cost and high efficiency. Recent research on aluminium based sorbents such as Al nanoclusters, polyaluminum chloride, polyaluminum granulate, and gibbsite have revealed that As(v) forms bidentate-binuclear complexes. Al nanoclusters appear to be very promising. | 14 |
| 4. | Biogenic jarosite | Jarosite is essentially a basic hydrous sulphate of potassium and iron, which is usually produced as a byproduct during refining of zinc and found in acid mine drainage systems. Arsenate has been found to adsorb on jarosites and biogenic jarosites are reported to have better adsorption properties as compared to their synthetic counterparts. | 15 |
| 5. | Biopolymer-reinforced synthetic granular nanocomposite technology | These filters consist of granular composites, composed of metal oxyhydroxide–chitosan nanostructures and they combine the dual effect of a constant release of silver ions in water (antimicrobial property) and the ability to remove toxic species like arsenic from drinking water. These water purifiers are point of use devices that can function without electricity and are affordable (US$2.5 per year per family). | 16 |

**Table 5.1** (*Continued*)

| S. no. | Technology | Salient features | Reference |
|---|---|---|---|
| 6. | Photocatalytic oxidation with molybdenum oxide supported on Titania with subsequent adsorption of Arsenic(v) by $\gamma$-$Al_2O_3$ | In conventional processes, arsenite is oxidized to arsenate and subsequently filtered. However, this leads to oxidation byproducts. Therefore, a titanium dioxide based photocatalyst has been used as an alternative. Yet, the arsenate formed invariably accumulates on the catalyst surface. Combination of molybdenum oxide supported on Titania with subsequent adsorption of Arsenic(v) by $\gamma$-$Al_2O_3$ could overcome this drawback successfully. The arsenic is completely released and filtered downstream. This improved the overall efficiency of the process and reduced the operating cost as the catalyst did not need frequent reactivation. | 17 |

useful as point of use devices. More research and efforts are required to make them applicable on a larger scale. Nevertheless, these methods are very promising and hold tremendous potential for developing countries.

### 5.1.4 Arsenic Removal by Subterranean Arsenic Removal (SAR) Technology in West Bengal, India

#### *5.1.4.1 Background*

Kasimpore is a village that is situated around 25 km from Kolkata in West Bengal, India. Water in this village is reported to be contaminated with elevated levels of arsenic. The villagers heavily depend on shallow wells and tube wells for drinking purposes but nearly 70% of these tube wells had high levels of arsenic. Arsenic levels in water were reported to be above 50 $\mu g\,L^{-1}$, which is much above the 10 $\mu g\,L^{-1}$ WHO guideline value. The alarming health risks due to elevated levels of arsenic in drinking water especially in rural areas were already emphasized in the previous sections. To combat the deleterious effect of arsenic laden drinking water, a team of European and Indian scientists used the SAR technology to remove arsenic from potable water.

#### *5.1.4.2 What Is Subterranean Arsenic Removal (SAR) Technology?*

Among the conventional arsenic removal methods, activated alumina filtration has been most frequently employed on a community scale.

This technique entails pumping of arsenic contaminated water on the earth's surface and then passing it through activated alumina filter beds that adsorb the arsenic. However, one of the major drawbacks of this technique is clogging of these filters and the need for replacement of these filter beds, which is often expensive for the rural population. The SAR technology has addressed this issue by converting underground aquifers into a natural biochemical reactor that removes arsenic along with other contaminants such as iron and manganese, which are often present in water, by the process of adsorption. Therefore, no filter beds are needed, and the naturally present chemo autotropic microorganisms help in the acceleration of the oxidation processes. The autocatalytic effect of the oxidation products that are formed also aids in further augmentation of the oxidation process.

Thus, in the SAR technology, the oxidation and filtration processes that are conventionally carried out on the ground surface occur underground within the aquifer itself. It requires a simple set-up that is shown in Figure 5.1. It essentially consists of a tube well that is dug into the earth to a depth of about 37 m. Underground water from the aquifer is pumped by means of a submersible pump through the tube well into a storage tank that is kept above ground level. Aeration of water in the storage tank using simple plastic shower heads results in oxygenation of the water (dissolved oxygen level up to 6 mg $l^{-1}$) and approximately 15 to 20% of this water is returned to the aquifer at the same depth under gravity. The remaining water from the storage tank is supplied to the end users as arsenic free water.

It is important to note that the set-up is operated in such a way that when delivering water from the aquifer to the storage tank, adsorption of Fe(II), As(III) and Mn(II) occurs on the sand grains in the aquifer. This is followed by a rest period, and subsequently, oxygen rich water is made to infiltrate the aquifer that results in the formation of an oxidizing zone deep in the aquifer.

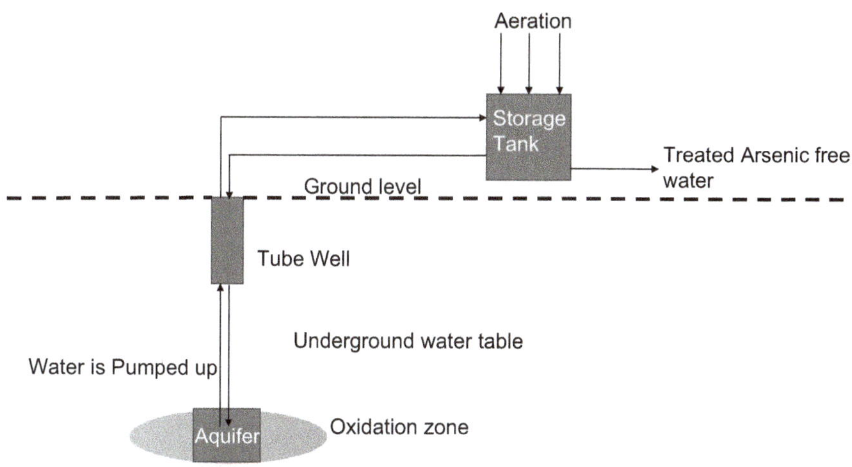


**Figure 5.1** The SAR Technology set-up.

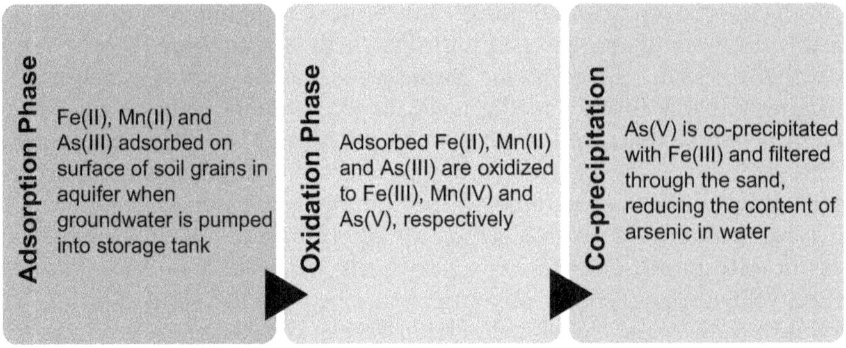


**Figure 5.2** The process of Subterranean Arsenic Removal (SAR) Technology.

Here, oxidation of the adsorbed impurities occurs and the dissolved oxygen in the aerated water converts Fe(II) to Fe(III), As(III) to As(V) and Mn(II) to Mn(IV). This is also accompanied by the co-precipitation of As(V) with Fe(III). These flocs are then easily filtered through the sand and the filtered water has arsenic that is well within the prescribed limits of 10 $\mu g L^{-1}$ (Figure 5.2). This is one of the greatest advantages of this method as the arsenic and iron that are trapped in the sand constitute a negligible volume of the total volume being handled. Thus, there is no sludge handling and disposal issues. Apart from this, the SAR method offers several advantages as compared to the conventional methods for arsenic removal.

## 5.1.5 Merits and Demerits of SAR Technology

This technique is a sustainable process well suited to developing countries as it is free from chemicals and waste, and requires low maintenance, which can be easily carried out by the local technicians. It is also reported to be less dependent on the source water quality. However, the iron content in water has a big role to play. The SAR process becomes more effective when the iron content in water increases, as the arsenic gets adsorbed on the soil particles coated with Fe(III) as explained previously. It has been found to reduce arsenic content in the range of 100 to 250 $\mu g l^{-1}$ to the permissible levels of 10 $\mu g l^{-1}$. Particularly at the Kasimpore plant, the SAR technology has effectively delivered low arsenic (<10 $\mu g l^{-1}$) water since inception. The salient features of this project are shown in Table 5.2.

The SAR technology can handle large volumes and is an economical process having low capital cost and minimal operational expenditure with a typical production cost of 1 US$ for every 10 000 litres. Around six plants based on this technology are now in operation in the rural locations of West Bengal, India. The typical capacity of these plants is between 3000 to 4000 litres per cycle and the arsenic levels have been successfully reduced from the initial concentration of 150 to 280 $\mu g l^{-1}$ to the permissible levels

**Table 5.2** Highlights of SAR Technology at Kasimpore, West Bengal, India.

| Capacity of the plant | Arsenic levels before treatment | Arsenic levels after treatment | Cost of treatment | Total population benefitted |
|---|---|---|---|---|
| About 3000 to 4000 l per day | $>50\ \mu g\, l^{-1}$ (usually around 91 to 282 $\mu g\, L^{-1}$) | $<10\ \mu g\, l^{-1}$[a] | 1 US$ for every 10 000 litres | Approximately 5000 people |

[a]Permissible limit of As in water as per WHO guidelines is 10 $\mu g\, l^{-1}$.

of 10 $\mu g\, l^{-1}$. These plants are providing water to six rural communities each with a total population of more than 5000.

The main disadvantage of the system appears to be the time taken for the system to stabilize due to the slow kinetics of the oxidation process. Moreover, it is reported that when the oxidation of the aquifer is not controlled, it can lead to As and Fe precipitation rather than adsorption. This is not desirable as it may lead to a release of As in the water subsequently. The sudden change in the redox potential can cause deleterious effects on the bacterial population thereby making the SAR process ineffective. Additionally, the spray systems employed should work effectively to bring about aeration and only submersible pumps (hand pumps are not adequate) are suited for this process and electric power is a requirement to operate the same, which adds to the cost.

Overall, the SAR technology has been effective in delivering water with low levels of arsenic and the success of this technology has resulted in its application in other parts of the developing world such as in Malaysia and Cambodia. Presently, the group reports that several projects are in the pipeline with the help of collaboration with different agencies to bring this technology to the people on a larger scale.

## 5.1.6 Conclusions

Given the critical condition of people exposed to elevated levels of arsenic in West Bengal and the severe health risks involved, it is imperative to find a simple yet efficient and economical solution. SAR technology appears to be one such methodology that has proven its effectiveness in all the plants that have been set up in West Bengal. The technology has some demerits that have been discussed in the preceding sections and mainly include the sensitivity to pH, presence of other competing pollutants, and the selectivity and surface characteristics of the adsorbent. However, the simplicity of the process that does not necessitate elaborate equipment and skilled labor makes it apt for rural communities as the locals can be trained to operate the system. SAR being an *in-situ* process also reduces waste generation and is environment-friendly. Major efforts from the experts to scale this even further are a promising factor for the generations to come that can drink and use water that has low arsenic levels.

## 5.2 Case Study 2 – *Moringa oleifera* – A Natural Way to Safe and Clean Drinking Water

### 5.2.1 Nature's Bounty for Water Treatment

For eons, humans have used various natural substances for cleaning water even before chemicals and other technologically advanced methods existed. Herbs and plant sources were used in different parts of the world for treating water and they are still prevalent in many parts of the globe. A detailed discussion on herbal approaches for water treatment has already been described in detail in Chapter 2 (Section 2.2.3). The key reason for selecting a herbal approach for a case study is mainly because plant material is indigenous to the region and is locally easily available. People living in such areas need not depend on the purchase of expensive chemicals or other modes of water treatment that may require electricity. Naturally available herbs offer a very good alternative to traditional treatment methods due to their several advantages.

Although many merits have been highlighted in Chapter 2 (Section 2.2.3), it is useful to recall some of them here. Herbs are varied in number and have different active ingredients, which can be in any part of the plant such as the roots, stems, leaves, seeds *etc.* These plants are abundantly grown in the equatorial regions and are hence locally available for the population in developing countries located geographically around the equator like North Africa, Maldives, Sudan, Congo, India, Indonesia, Columbia and Brazil. The active ingredients of these herbs have shown coagulating and antimicrobial properties due to which they have been extensively employed in water treatment.

Among many herbs, such as tulsi, tamarind, okra *etc.*, that have been used as water treatment options, *Moringa oleifera* (drumstick) appears to be amply used and researched as an option in rural areas. Most of the studies have focused on the use of *Moringa oleifera* for reduction of water turbidity. Some studies also point to the use of *Moringa* as an antimicrobial, which means that in addition to getting clear water, one can expect that the microbial load of the raw water will be substantially reduced. Most of these aspects have been sufficiently described in Chapter 2 (Section 2.2.3). It is interesting to note that most of the research on the use of *Moringa oleifera* for water treatment has reported investigation at the household or point of use level and many such studies have used model water (synthetically prepared turbid waters or/and the use of standard microbial species).

### 5.2.2 Toxicity Studies of *Moringa oleifera*

Natural herbs offer economic and viable options for water treatment in developing countries, however, they may pose a threat to human health because of the toxicity of these natural herbs and their active ingredients.

This has already been emphasized in the earlier chapters, but it is important to note that toxicity studies have been conducted in the past and such efforts are also underway in the present. Many a times there are different opinions expressed by scientists about the toxicity and the maximum concentration of the *Moringa* Seed dosage that can be used safely for water treatment purposes. For instance, early investigations in the year 1985 revealed that *Moringa oleifera* demonstrated toxic effects due to the active antimicrobial constituent 4(a-L-rhamnosyloxy)benzyl isothiocyanate. However, the studies revealed that the *Moringa oleifera* extract did not appear to cause human health issues at least in the concentrations generally used for water treatment purposes.[2]

Over the years, numerous attempts have been made to ascertain its harmfulness and in one such experiment performed in the year 2014, both the cytotoxicity and genotoxicity of *M. oleifera* were reported. Substantial cytotoxicity effects were observed when the powdered *M. oleifera* seeds were used in the concentration ranging from 1 to 50 $mg\,L^{-1}$. The cytotoxicity has been attributed to the presence of hydrophobic lipids in the *Moringa* seeds. The lethal concentrations (LC50 and LC90) of *M. oleifera* seeds were reported to be 8.5 $mg\,L^{-1}$ and 300 $mg\,L^{-1}$, respectively, and their genotoxicity was found to be equivalent to 8.3 mg of mitomycin C per 1.0 g of dry *M. oleifera* seed.[3] Thus, careful consideration is required when *Moringa oleifera* seeds either as seed extracts or as the purified active component are used for water purification.

There are numerous literature studies on *Moringa* and although these studies are important to understand the effectiveness of *Moringa oleifera* in turbidity reduction and its mechanism of inactivation of potential pathogens, they do not give a clear picture on full scale applications. Trials at pilot and full scale are reported only to a limited extent and therefore the following case study although not very recent has been chosen to examine the use of *Moringa oleifera* on a full-scale level.

### 5.2.3 *Moringa oleifera* as Coagulant for Water Treatment in Malawi, Africa[4]

#### *5.2.3.1 Background*

Malawi is a country in southeast Africa and it is one of the least developed countries in the world. Almost 85% of its population lives in rural areas and they heavily depend of agriculture for their economic development. Malawi has a rich source of water from various lakes, rivers, aquifers and wells and people here rely on these water sources for potable and domestic uses. The main issue is not the availability but the safety of these sources as many of these surface and groundwater sources are heavily contaminated by various chemical and microbial pollutants. This polluted water is often highly turbid, making it very difficult for the locals to access safe and clean water for drinking.

Consuming turbid water is not only aesthetically unacceptable but it also leads to the consumption of pathogenic microorganisms and toxic chemicals that can cause health risks among the people of Malawi. According to UNICEF, water borne diarrheal disease is one of the main reasons for deaths in children below 5 years in Malawi. Therefore, water treatment in Malawi is of prime significance. However, the economic conditions in this developing nation are very poor as most of the population depend on agricultural activities for their daily living. This clearly points out that the mechanisms of water treatment should be based on cost effective technologies that can be easily adopted in Malawi.

Among the plethora of water treatment techniques that have been tried and tested in developing nations, methodologies that are based on simple, effective and economic techniques are the ones that benefit the local population. Most of the water treatment studies in Malawi have reported methods to reduce the turbidity of the source waters and there are some investigations on the inactivation of microbes as well. Since turbidity is visually detected by the consumer, it becomes necessary to first reduce the turbidity to acceptable limits and then work on reducing the other pollutants. Often, reducing turbidity also results in the removal or reduction of these contaminants as well.

For removing turbidity in water, simple chemical treatment with alum (aluminium sulphate) or the use of slow sand filters has been preferred in such rural set-ups. In Malawi, treatment with alum is costly and requires skilled people for its application, and most of the time, the quality of the water is not known. In the case of sand filtration, there is also the initial cost and maintenance of the filters to prevent clogging. Therefore, natural ways to treat water in Malawi are highly recommended as they would be effective and economical as well. Use of plant-based coagulants is one such approach.

### *5.2.3.2 Why* Moringa oleifera*?*

It is interesting to note that among different plant materials that have been used as coagulants to treat turbidity, *Moringa oleifera* appears to be the plant of choice in Malawi. This could be because this plant is grown abundantly in Malawi and its ready availability and ease of use are advantageous to the rural population that depends on agriculture for their living. *Moringa oleifera* tree provides several products from different parts of the tree that can be used by the locals. For instance, the green pods, leaves and flowers are used in cooking flavorsome dishes, the wood is used as a fuel source and the tree is grown as a live fence too. The seeds contain almost 40% oil by weight and are therefore crushed to get vegetable oil that is used as cooking oil and in soap manufacture. The press cake residue remaining after oil extraction is used as a coagulant for water and waste water treatment.

The press cake is reported to contain the active ingredients that bring about coagulation and since the press cake is a byproduct of oil extraction, it is a very cost-effective alternative to chemical coagulants. This economic and viable option of using *Moringa oleifera* as a natural coagulant for treating water in Malawi is equally effective and cost effective when the seeds are used directly (not as a byproduct) as the tree cultivation in Malawi is abundant and even if the seeds are purchased from the local farmers, it would still be economical as compared to purchasing chemicals from other countries. Moreover, it would also encourage small scale farmers by giving them opportunity and money as well. A specific study based on the use of *Moringa oleifera* to reduce turbidity of water in Malawi is presented here.

### 5.2.3.3 *Full Scale Water Treatment with* Moringa oleifera

A full-scale water treatment was conducted at the Thyolo treatment works in southern Malawi with the support of the local Ministry of Works Water Department. The raw water used here was river water that had very high turbidity levels ranging between 270 to 380 NTU. The treatment consisted of various processes such as flocculation, clarification and rapid sand filtration, as shown in Figure 5.3. The main aim of the water treatment project was to replace the chemical coagulants such as alum that were routinely used with natural coagulants consisting of *Moringa oleifera*.

For these trials, *Moringa oleifera* was purchased from the local villagers as most of them planted these trees and therefore there were abundant seeds available. There is a general procedure followed for the preparation of *Moringa* seeds for water treatment. It essentially involves drying the seeds. For this, the seed pods are allowed to dry naturally on the trees before they are harvested, following which the seeds are shelled, crushed and sieved using traditional methods. Subsequently, dosing solutions are prepared, which are generally 1 to 3% solutions obtained by dissolving the crushed seed powder with water. The water-soluble proteins (active component) that are present are usually cationic in nature and act as polyelectrolytes that bring about coagulation when added to raw water.

The water treatment was conducted during the monsoons in the year 1992 when the source river water had very high levels of turbidity. The plant was

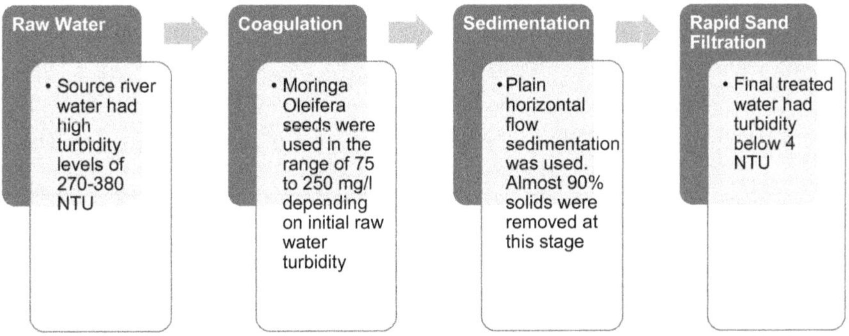


**Figure 5.3** Water treatment with *Moringa oleifera*.

operated at a flow rate of 60 $m^3 h^{-1}$. The treatment essentially involves dosing of *Moringa oleifera* solutions (75 to 250 mg $l^{-1}$) into the raw water with the help of centrifugal pumps, which is monitored using a rotameter. This results in the adsorption of the cationic polyelectrolyte present in the *Moringa* seeds followed by charge neutralization, which results in coagulation of the turbid raw water.

After the coagulation step, the solids can settle in a plane horizontal sedimentation tank. At this stage, almost 90% solid removal took place and a subsequent rapid sand filtration resulted in water that had a very low turbidity of >4 NTU. This was used as the final water that was supplied to the end users. Thus, a simple substitution of *Moringa oleifera* seed extract instead of alum not only reduced the chemical consumption and cost of the treatment plant but also resulted in pure and clean water that was aesthetically appealing, potentially safe and very economical for a developing country like Malawi.[5]

### 5.2.3.4 *Economics*

Here, it is important to note that the active component of *Moringa oleifera* seeds responsible for coagulation has been isolated and purified by many experts. Interestingly, it was observed that when the purified active ingredient is used, the dosage required is less compared to the seed extract required. However, this implies that purification steps are needed to obtain the pure active ingredient, and this leads to numerous treatment steps and the corresponding cost augmentation of the process. Cost effectiveness is the main reason for the use of direct seed extract for water treatment in this case study.

The salient features of this trial are presented in Table 5.3 and it can be observed that *Moringa oleifera* seeds were purchased from the local farmers who planted these trees. The cost in Malawi Kwacha (MK) per 1000 $m^3$ of water treated was 75 (at the time of this study in March 1993, MK 10.07 was equal to £1 sterling). Fascinatingly, an economic analysis carried out by the experts in the Malawi context revealed that the press cake (residue obtained after oil extraction from *Moringa oleifera* seeds) may be obtained at zero net cost by the water utilities department, which could make this treatment scheme even more economically attractive.[5]

**Table 5.3** Highlights of full scale water treatment with *Moringa oleifera* at Thyolo treatment works, Malawi (Sutherland *et al.*, 1994).[a]

| Capacity | Inlet turbidity | Final turbidity | *Moringa oleifera* dosage | Cost in Malawi Kwacha (MK) per 1000 $m^3$ of water treated |
|---|---|---|---|---|
| 60 $m^3 h^{-1}$ | 270–380 NTU | <4 NTU | 75–250 mg $l^{-1}$ | 75 |

[a]*Moringa* seeds purchased from local farmers and MK 10.07 = £1 sterling in March 1993 when this trial was done.

#### *5.2.3.5 Conclusions*

Thus, *Moringa oleifera* appears to be a good choice for water treatment, particularly to reduce turbidity of highly turbid waters. It is also a promising alternative to traditional coagulants such as alum, which are often an expensive choice for the rural population. When used in combination with chemicals, *Moringa oleifera* is found to substantially reduce the quantity of chemicals needed to reduce turbidity and hence the corresponding cost. In tropical areas such as Malawi as described in the case study, *Moringa oleifera* is abundantly grown. Therefore, it becomes a plant of choice for water treatment as it is easily available, and it also provides an avenue of job opportunity for the local villagers. However, the toxicity aspects of using any natural plant source cannot be overlooked. This becomes especially important since consuming water treated with natural herbs like *Moringa oleifera* will lead to a direct introduction of these plant components into the human body. Therefore, further studies are required in this area as the prevailing reports often differ in their conclusions. Overall, *Moringa oleifera* appears to be a promising natural plant for water treatment in developing countries.

## 5.3 Case Study 3 – Rain Water Harvesting: A Natural Way to Water Conservation

### 5.3.1 Why Rain Water Harvesting?

Water scarcity is an issue faced by all countries worldwide and it is particularly challenging for the people living in developing countries. This may be attributed to the limited access to water for drinking and other domestic uses. Most water sources such as surface water and ground water are slowly getting depleted due to many factors such as overuse, increasing population and even environmental factors such as global warming. Therefore, finding other sources of water that can be used for the domestic needs of the people is of paramount importance globally. This is especially significant when applied to the rural scenarios of developing countries.

Conservation of water by rain water harvesting appears to be a promising solution to combat this water scarcity. Rain water harvesting can be easily carried out in most developing countries that receive high rainfall at least in the monsoon months. Since it is a natural process, it can be conveniently used for the benefit of humankind and at the same time save this precious resource that is otherwise getting depleted slowly. Rain water harvesting has been elaborated in Chapter 3 where a detailed discussion on the need for harvesting rain, system design, treatment of harvested water, health aspects and economics of the process has been presented. Here, a brief overview of the same with an emphasis on the current studies on RWH is presented followed by a case study.

### 5.3.2 Current Research on Rain Water Harvesting

It is fascinating to note that over the last few decades, numerous initiatives have been undertaken to study rain water harvesting and its application in developing countries. Researchers have focused on various aspects of this technique to make it more efficient, safe and economic for the rural population. Some of these studies have been described in Table 5.4. Among the myriad investigations that are in progress, the selection of appropriate roof top materials, proper design and sizing of the catchment area, estimating the volume of storage tank needed and cost-effective materials for construction of tanks, pipes and fittings are some of the issues that have been and are being addressed. For instance, the average monthly rainfall is usually employed in the design of a RWH system and this may lead to errors. Therefore, a RWH model was developed that was based on daily rainfall and this gave very accurate performance predictions of the RWH systems.[6]

Most importantly, the quality of the harvested rain water has been tested and various ways of treating the harvested water to ensure it is free from chemical and biological pollutants are also currently being explored by various experts in this field. Treatment of harvested rain water has already been elaborated in Chapter 3. However, this area continues to be an important area where novel techniques and innovative disinfection practices are being studied by experts worldwide. Very recently, the use of a pilot scale solar photocatalytic fixed bed tubular reactor for treating harvested rain water has been demonstrated. The authors report that the rain water was completely disinfected and there was a complete removal of chemical oxygen (COD) as well with 2 h of solar treatment.[7]

Last but not least, the cost of rain water harvesting systems and their affordability are crucial factors that will determine its implementation in any locality. This is even more important if it is a developing country with low economic conditions. Therefore, some efforts are also being concentrated on the economic aspects so that RWH can be easily adopted by rural communities in low socio-economic groups. One such interesting study states that the success of RWH systems in developed countries like Australia can be effectively used in developing countries such as Kenya and an economic analysis tool called ERain has been presented. The authors use ERain to combine daily performance analysis of RWH systems with life cycle cost analysis for use in economic evaluation. The results indicate that there is a need to innovate and reduce the capital costs as well as on-going costs of RWH systems. Additionally, it is indicated that people must be encouraged to install larger tanks to increase the water security. For this, the authors suggest the use of a rebate that matches the tank size.[8]

Thus, efforts towards efficient and economic ways of implementing RWH in developing countries are underway. As an example, a case study based on RWH and its implementation in India is presented below. India receives

**Table 5.4** A few examples of studies conducted on rain water harvesting in developing countries globally.

| S. no. | Developing country | Studies on rain water harvesting | Reference |
|---|---|---|---|
| 1. | Zambia | Rooftop rain water harvesting was successfully implemented with local materials and available skill. Ferrocement tanks that had been estimated to cost US$470 for 10 $m^3$ size were selected. Although the overall quality of harvested water was within the WHO limits, the presence of debris and taste issues could not be ruled out. | 18 |
| 2. | Edo State, Nigeria | A statistical study conducted revealed that almost 80% of the households practiced RWH, which was mostly used for drinking purposes. Rooftops of the houses were used as the catchment area. Bacteriological quality of the harvested water was poor due to the presence of coliforms. It was concluded that proper design, maintenance and health education to prevent the spread of diseases were needed for correct implementation of RWH. | 19 |
| 3. | North Africa | A new method of finding the reservoir volume as a function of rooftop area, number of people living in a house and rainfall distribution was investigated. The findings showed that more than 30% savings could be made on total cost of water and around 50% savings on water volume at source. | 20 |
| 4. | Mexico (Rural communities located in the San Miguel de Allende Municipality) | In this study, the harvested rain water was treated with silver ions as a disinfectant to improve its bacteriological quality. The device was essentially a pair of silver electrodes that were charged by an alternating DC voltage that was used along with filtration. Initial total coliform counts of about 1000 CFU/100 ml reduced to zero after treatment and some amount of silver remained in the distribution systems that was beneficial to prevent recontamination. | 21 |
| 5. | Hanoi, Vietnam | The harvested rain water was found to be of relatively good quality compared to surface and ground waters. UV treatment of harvested water resulted in complete inactivation of total coliforms and *E. coli*. The authors stressed the need for rain water quality guidelines especially for developing countries. | 22 |

good rainfall in most parts of the country, especially in the monsoon months spanning between June and September. This makes rain water harvesting a very viable option in India. Moreover, the quality of harvested rain water is often relatively better than the existing surface and groundwater sources that are usually employed for potable purposes. Even if the contaminant levels are more than acceptable limits, appropriate disinfection methods as discussed in Chapter 3 can be easily used to get safe and clean drinking water. The case study selected addresses the design aspects of a RWH system and the economics of the process.

### 5.3.3 Water Conservation by Rain Water Harvesting in Dhule city, Maharashtra, India

#### *5.3.3.1 Background*

Dhule town is situated near Chalisgoan in the state of Maharashtra in the western regions of India. It is located near the Panzara river and its outskirts constitute a total area of about 26.68 $km^2$. Although many parts of Dhule are slowly developing and constitute peri urban characteristics, several areas of Dhule lack basic facilities such as water supply. There are numerous ground water sources available for domestic and agricultural use in Dhule. Over a period, these sources have been slowly depleted as the water levels have gone down. Consequently, efforts to recharge the ground water table and to find alternative water sources are the need of the hour. One solution to this is the use of rain water harvesting, which is an effective way of water conservation. The water collected by harvesting rain from the rooftops of houses can provide an alternative source of water for domestic use including drinking. The surplus collected in this manner can also be used to recharge the ground water table.

This approach is well suited to a city like Dhule because it receives very good rainfall like most parts of India during the monsoon months between June to September. The rainfall data recorded in Dhule for a period of 10 years between 1996 and 2006 clearly point out that around 117.38 to 124.10 mm of average rainfall is available in the monsoon months and the maximum rainfall is in the month of July (Table 5.5). Before and after the monsoon months (June to September), the rainfall was found to reduce gradually. Thus, the period between June and September can be used advantageously to collect the rain water and use it for domestic needs and to recharge the groundwater table.

#### *5.3.3.2 Methodology of Rain Water Harvesting and Its Cost Analysis*

In this study, around 50 houses were selected where each house had a different roof top area. The rain water harvesting potential for different rooftop areas was calculated. In general, the rooftop areas of the houses were in the range of 50 $m^2$ to 250 $m^2$. In Dhule, when there is less rainfall or during the non-monsoon periods of the year, the use of only Domestic Rain Water Harvesting (DWRH) does not lead to fulfilling the demand for water. Additionally, it is evident that the rooftop areas of the houses play a huge role in the collection of rain water and it was observed that only for rooftop areas above 125 $m^2$ can the gross water demand be met during the four monsoon months between June and September. Therefore, rooftop area values of 150 $m^2$ (an example of a rooftop area above 125 $m^2$) have been highlighted in Table 5.5. The size of the water tanks needed to fulfill drinking and cooking demands at the rate of 8.00 litres per capita per day (LPCD) was calculated as follows:

$$V = -16.8\,A + 8502$$

where $V$ = volume of the water tank in litres and $A$ = rooftop area in $m^2$.

The overall methodology for estimating the tank size needed for rain water harvesting and the excess water available for ground water recharge was planned as described in Figure 5.4. Firstly, information on per capita water demand of a household and the number of dwelling members were utilized to arrive at the monthly cumulative demand of water. Secondly, information on the monthly average rainfall in Dhule and the rooftop area runoff coefficient were applied to arrive at the cumulative rain water harvesting potential. Both these generated data were then employed to estimate the volume and tank size required for rain water harvesting and the quantity of water available for groundwater recharge.

These data were collected for all the months in a year. It was observed that during the monsoon months, especially in the month of July, the average availability of rain water for harvesting was maximum. Therefore, rain water harvesting related data including the cost element for the month of July are shown in Table 5.5. The costs of different tank sizes needed for different rooftop areas were found out based on DSR-2007–08 of Maharashtra Jeevan Pradhikaran, Nashik Region. The Maharashtra Jeevan Pradhikaran was formerly known as the Maharashtra Water Supply and Sewerage Board (MWSSB) and was primarily founded for the rapid development and proper regularization of water supply and sewerage services in the State of Maharashtra, India. District Schedule Rate (DSR) is a document that lists the pricing of labour, plant, hiring charges *etc.* The general equation used to calculate the cost of a water tank and the rooftop area is:

$$C_T = 1.0022 \times V + 4332.80$$

where $C_T$ = Cost of Water Tank in Indian Rupees and $V$ = volume of the water tank in litres.

In this study, the quantity of pipes, fittings and other materials associated with RWH and cost of all materials needed were measured in such a way that the water that was harvested could be either used for domestic purposes or

**Table 5.5** Highlights of rain water harvesting in the month of July in Dhule city, Maharashtra, India.

| Rooftop area | Average Rainfall in July (1996 to 2006) | % Total Water Demand fulfilled by DRWH | Quantity of Annual Harvested Water For drinking | Total capital Cost | Gross Annual Cost | Cost of Water from DWRH | Cost of Water from DWRH (in tank) + 20% useable from GW Recharge |
|---|---|---|---|---|---|---|---|
| 150 $m^2$ | 124.10 mm | 127.60 | 14 600 L | INR 20 225.72 | INR 1677.52 | INR 114.90 $kL^{-1}$ | INR 63.85 $kL^{-1}$ |

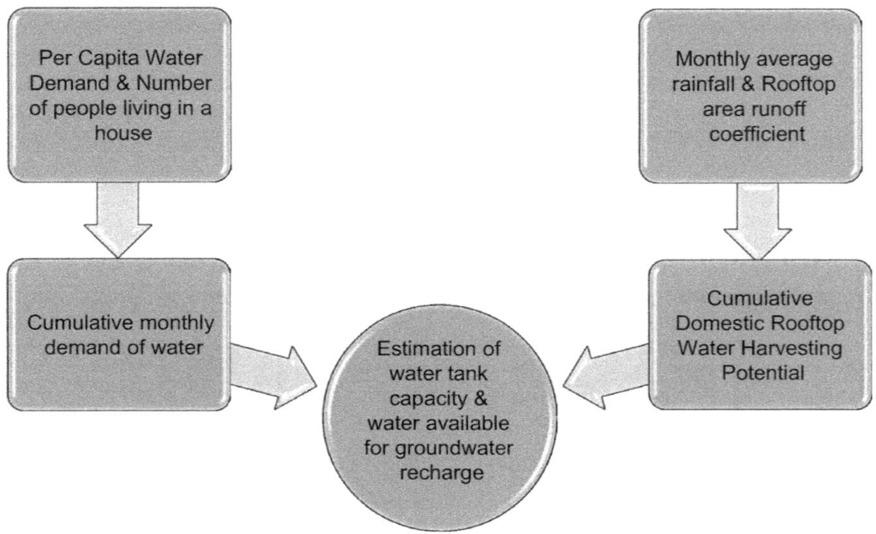


**Figure 5.4** Rain Water Harvesting – Process of Assessment of Tank Size.

as groundwater recharge. Alternatively, the harvested water was also considered for both uses. The general formula for estimating the cost of a DRWH system for houses in Dhule with different rooftop areas was as follows:

$$C_{\mathrm{D}} = 3265 \cdot \mathrm{Ln}\,(A) - 6462,$$

where $C_{\mathrm{D}}$ = Cost of DRWH system in Indian Rupees and $A$ = Rooftop area in $m^2$.

For determining the unit cost of harvested water, some basic assumptions have been made such as considering 30 years of serviceable life of the RWH systems including the storage water tanks. The rate of interest was considered as 7% and the salvage value of the RWH systems was assumed to be 25%. Moreover, since maintenance and periodic repair are often needed for RWH systems, the annual maintenance and repair costs were considered as 0.5% of capital cost. The cost estimates are shown in Table 5.5., and since maximum rainfall was obtained in the month of July, only values related to the average rainfall in the month of July have been highlighted.

The cost of water from DRWH is around INR 115 $kL^{-1}$ and the market cost of water in Dhule town at the time of study was reported as INR 125 $kL^{-1}$. Therefore, the cost of harvesting rain water is less than the market cost of water. Additionally, a portion of the harvested rain water was used for groundwater recharging, which can be subsequently used for other means. This means that the cost of water from DRWH could be reduced further (almost INR 64 $kL^{-1}$), as shown in Table 5.5, which is almost half the market value.[9]

### *5.3.3.3 Conclusions*

Domestic rain water harvesting in Dhule town was very effective as it provided the people with water for drinking and cooking in a very simple, effective and economic way. The methodology used for designing the tank and estimating the cost of the DWRH can be extended to other areas too. The mean rainfall was considered in this study and it may be beneficial to consider the median value of rainfall for a more accurate value. Moreover, the amount of rainfall was considered in this study, but its intensity was not recorded. Inclusion of such information can make the designing of rain water harvesting systems even better. Overall, any house when planned at the time of construction to include provisions for rain water harvesting such as proper roof tops, drainage pipes, storage tanks and provisions for groundwater recharge can lead to efficient and economical rain water harvesting systems. Thus, nature's gift can be conserved efficiently and used in the appropriated manner.

## 5.4 Case Study 4 – Waste Water Reuse for Direct/Indirect Potable Use in Developing Countries

### 5.4.1 Need for Water Reuse

The last section on rain water harvesting and the case study based on it clearly points out that conserving water can play a huge role in protecting our water resources. Yet another significant way of saving water is by waste water reuse and this area has been addressed as a separate chapter (Chapter 4) in this book. A detailed discussion on the reuse of waste water with an emphasis on indirect and direct potable use in developing countries has already been given. Waste water reuse has become increasingly important in recent years and it also appears to be a feasible solution because waste water generated by different industries can be effectively treated and reused for different purposes. In developing countries, industrialization is also increasing consistently and therefore there is an augmenting opportunity for waste water reuse in such economies. Asian countries such as India have a huge chance of utilizing treated waste water to meet the indirect and direct potable needs of the population.

### 5.4.2 Waste Water Reuse in Developing Countries Globally

Innumerable projects for reusing waste water have been undertaken in several parts worldwide to address the increasing global water demand. Although many of these projects have been for non-potable use, several initiatives have also been undertaken for generating drinking water either for direct or indirect potable reuse. Most of the discussions related to this have already been highlighted in Chapter 4. Many interesting projects around the world, especially in developing nations, are being implemented to reuse treated waste water for different end uses depending upon the degree of treatment, the quality of treated waste water and the needs of the consumer.

For instance, in the rural wadi valleys of Saudi Arabia and other Middle east areas and North Africa, water scarcity for drinking as well as irrigation is a major challenge. An effective solution for this issue could be either the use of desalinated seawater or treated waste water, which could be transported by means of a managed aquifer recharge system. An economic analysis revealed that the cost per cubic meter for supplying desalinated water was US\$2–5 $m^{-3}$ plus the conveyance cost. On the other hand, the cost of treated waste water conveyed *via* an MAR system was US\$0–0.5 $m^{-3}$ plus the transport cost. This study shows that the economic loss caused by waste water discharge to the water bodies can be higher than the overall cost to construct an MAR system including the transport pipelines and the operational cost of waste water reuse in a rural set-up.

The treated waste water can be reused for direct irrigational purposes or indirectly for potable use. This is a very good way to conserve water that would otherwise have been discharged into the marine environment leading to major pollution and even coral reef destruction. Such situations are commonly seen in Arabia and other such areas situated in arid lands worldwide. In areas where waste water is being discharged with no economic benefit and where alternative sources of water are very expensive, MAR systems have the potential for exploration thus leading to water conservation in a safe and cost-effective manner.[10]

According to the International Water Association (IWA), many cities in developing countries across the globe have started using water reuse technologies. In a very fascinating waste water report from 2018, the IWA reports on eight different cities from developing countries where the existing and future water challenges are very intensely felt. Different ways by which these cities have embraced water reuse options are presented in Table 5.6. It is interesting to note that these cities from various developing nations have made a stupendous effort towards conservation of water by way of waste water reuse. From a meager 5% to a huge level of more than 90% of treated waste water reuse can be seen in these examples.[11]

The main driver in all the cases is the scarcity of water, and various stake holders prioritizing water conservation by means of using treated waste water has led to an augmentation of this effort. In most of the cases, this drive has led to the construction of waste water treatment systems thereby generating business opportunities in this sector and revenue generation for the government. Yet another major benefit has been the reduced levels of pollution by decreasing the flow of waste waters into the water bodies and thereby protecting the marine environment. The huge amount of energy savings by using energy from biogas is an attractive incentive to these regions.

Most of the treated waste water in these cities is being used for agricultural, irrigation and environmental purposes. In a few instances, it is also used for domestic needs. Although direct potable use may not be very prevalent currently, it is important to note that use of treated waste water for indirect potable and non-potable uses makes a huge amount of water available for drinking purposes, thus reducing the demand for potable water. Nevertheless, it is a leap forward towards the goal of potable reuse in the coming future.

One of the major bottlenecks in using waste water reuse systems is an appropriate economic cost benefit analysis of the system. Often, the cost of construction, operation and maintenance and quantifiable benefits are all that are considered in such projects. Due to this, many waste water projects may be undervalued and hence may not be put into operation. Therefore, over the last few years, attempts have been made to evaluate the true benefits by trying to quantify the non-monetizable benefits. This has resulted in using the benefit to cost ratio as an evaluation tool to ascertain the economic feasibility of waste water treatment projects. One such case study is presented in the following section.

**Table 5.6** Waste water reuse in cities located in developing countries across the globe (IWA waste water report 2018).

| S. no. | City in developing country | Waste water reuse | Capacity | Advantages |
|---|---|---|---|---|
| 1. | Aqaba city in Jordan (Middle east) | Reuse of reclaimed water covers 30% of the city's water demand. | 90% of the waste water is treated, equaling 31 000 $m^3$ per day. | The investment in waste water treatment and reuse pays off in terms of tourism, public health and overall well-being of the residents and generates more than 4 Million US$ in income for the AWC. |
| 2. | Bangkok city in Thailand (Southeast Asia) | Sludge is collected, treated and sold as fertilizer. The production of energy from sludge treatment is currently under investigation. | 40% of the waste water is being treated to produce 1.3B L per day and about 5% of treated waste water is being used currently. | Creation of new markets and generation of income for businesses<br>Substantial pollution reduction |
| 3. | Beijing city in China (East Asia) | Recycled a quarter of domestic waste water and improved treatment of up to 85% of industrial waste water discharge. | Around 88% waste water treatment results in 4.4B L per day.<br>Reused water is used for:<br>Agricultural irrigation = 47%<br>Environmental uses = 30%<br>Industrial uses = 20%. | Improvement in the water quality of the Liangshui River is expected to increase to 100% in 2018. |
| 4. | Chennai city in India (South Asia) | It is the first Indian metropolitan area to achieve 100% sewage collection. | Around 70% waste water treatment results in 769 M L per day<br>Percentage of treated waste water used:<br>City's Water Demand = 15%<br>Sold to Industries = 8%<br>Domestic water needs in newly built houses = 40.7% | Reduction in the GHG emissions and electricity consumption from the city grid.<br>Generation of new markets for waste water treatment manufacturers and businesses |

| | | | | |
|---|---|---|---|---|
| 5. | Durban city in South Africa (Country in Southern Africa) | 23% of the city's treated waste water is reused by local industries in their production processes. | 108 M L per day of treated waste water is generated of which 44% is reused currently | Recycling effluent has reduced the demand for potable water by 7% and reduced the quantity of effluent directly discharged into the environment by 10%. |
| 6. | Kampala city in Uganda (Africa) | Creation of call centers for septic tank emptying and setting up decentralized sanitation systems. | 87 M L per day of treated waste water is generated, and all of this is currently reused. | 15% of National Water and Sewerage Corporation's (NWSC) income is generated from sewerage services.<br>Alleviation of pollution in Lake Victoria<br>GHG emissions have also been reduced through the use of energy from biogas. |
| 7. | Lima city in Peru (South America) | Lima's Water and Sewer Company SEDAPAL<br>operates 21 WWTPs with a total capacity of 2.775 $m^3 s^{-1}$. | About 15% waste water is treated to generate 240 M L per day of which about 5% is currently being reused.<br>Reuse of 3.5 $m^3 s^{-1}$ of treated waste water is used to irrigate 3 400 hectares of parks and gardens. | Increased ownership and responsibility towards reusing waste water among all stakeholders.<br>Waste water reuse is incentivized through the creation of a legal framework and enabling environment. |
| 8. | Manila, the capital of the Philippines (Southeast Asia) | More than 58 decentralized treatment plants were constructed. | 510 M L per day of treated waste water is generated. | 100% coverage and safely managed/reuse of waste water and sludge by 2028 is proposed.<br>Reduced pollutant loads to the environment and regenerated key resources. |

## 5.4.3 Water Reuse in a School, Madhya Pradesh, India

### 5.4.3.1 *Background*

Madhya Pradesh is a centrally located state in India that spans 308 245 $km^2$ and has a population of around 28 928 245 (in the year 2008 reported in the case study). In many parts of this state, waste water reuse facilities are not commonly available mainly due to inadequate infrastructure amenities and a lack of proper waste water drainage systems. However, efforts are underway in many parts of the country to install greywater reuse systems. This study was mainly undertaken by the authors to ascertain the feasibility of water reuse in schools of Madhya Pradesh, which are around 3500 in number. A boarding school in the Ganganagar region, District Dhar of Madhya Pradesh, which comprises about 300 students, was assessed for the feasibility of waste water reuse. The results obtained for this school could be inferred for the remaining schools in the state.

### 5.4.3.2 *Waste Water Reuse System*

In this case study, the waste water essentially consisted of the waste water coming from the use of showers and basins only; the kitchen and toilet waste was not used. The waste water generated was subjected to treatment that consisted of three steps (Figure 5.5). In the primary treatment, the waste water was subjected to absorption using a synthetic sponge followed by sedimentation in a sedimentation tank that was baffled or graded. The secondary treatment consisted of filtration using gravel of size between 10 and 60 mm and sand filtration. Lastly, in the tertiary treatment process, the effluent was treated using aeration and chlorination, and subsequently, the treated waste water was pumped to an overhead tank. This treated waste water was mainly used for toilet flushing in the school, cleaning of school floors and for small scale irrigation processes.

The technical feasibility of this waste water reuse system was assessed by The National Environmental Engineering Research Institute (NEERI). NEERI is a research institute created and funded by the Government of India to focus on water supply, sewage disposal, communicable diseases and to some extent on industrial pollution. The performance of the system as well as the quality of water at the inlet and outlet of the treatment system was

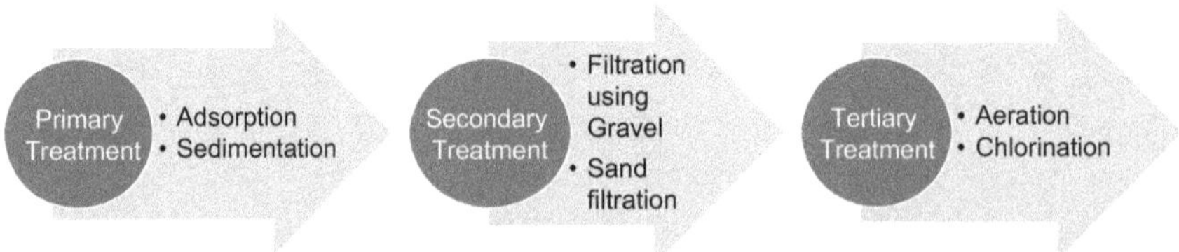


**Figure 5.5** Waste water treatment for reuse in a school, Madhya Pradesh, India.

monitored and found to be satisfactory. Moreover, the acceptance of the community, especially the school children, was studied, and it was found that the waste water reuse system was highly favored.

### 5.4.3.3 Economics

The cost of the waste water reuse system is a very important factor to be considered for its successful implementation. Therefore, a cost benefit analysis was carried out for the reuse system to ascertain the economic feasibility. Among the several factors that were considered, the internal benefits were one of the important aspects. This was obtained by subtracting the internal cost from the internal income, which is as follows:

$$\begin{aligned}\text{Internal benefit}\\ &= \text{internal income (obtained by allocating rate of tankered water}\\ &\quad \text{to treated greywater)}\\ &\quad - \text{internal cost (investment cost} + \text{operating cost} + \text{maintenance cost)}\end{aligned}$$

The cost of construction of this waste water reuse system was reported as INR 50 300 and the annualized cost consisted of the interest on capital cost amounting to INR 6036 and annual operating and maintenance cost of INR 5725, as shown in Table 5.7. The annualized benefits included several factors such as availability of greywater, evasion of water infrastructure, and environmental and health benefits. Each of these aspects play a huge role in the overall benefits of the waste water reuse system and lead to substantial savings in cost, as shown in Table 5.7.

Availability of waste water for reuse was around 4000 to 6000 liters per day on the school campus. The school depended heavily on tinkered water that had to be purchased from the local vendors at a rate of INR 30 for 1000 liters. In this case study, the value of waste water available for reuse was calculated on the basis of the tinkered water saved. In other words, the cost of treated waste water was the same as the cost of tankered water (INR 30/1000 L). Thus, the annual cost of waste water available was calculated as follows:

$$\begin{aligned}\text{Cost of available waste water for reuse}\\ &= \text{INR } 30/1000\,\text{L} \times 4000\,\text{L per day (considering the lowerside of availability)}\\ &= \text{INR } 120 \text{ per day}\end{aligned}$$

The study considers around 250 working days for the school and therefore the cost would be INR 30 000 for 250 days.

Similarly, the possibility of waste water reuse in the school premises will lead to the avoidance of water infrastructure that would be needed for a conventional fresh water source. In this case, a deep borewell would have

**Table 5.7** Highlights of the water reuse case study conducted in a school, Ganganager, Madhya Pradesh.

| Total number of students | Water demand | Waste water generated per day | Construction cost of greywater treatment and reuse system | Annualized costs: Interest on capital cost of waste water reuse system | Annualized costs: Annual operating and maintenance costs | Annualized benefits |
|---|---|---|---|---|---|---|
| 300 | 10 000 L School period: July 1 to April 30. | 4000 to 6000 L | 50 300 INR[a] | INR 6036 (interest rate @ 12% per annum) | 5725 INR[a] | Availability of greywater = INR 30 000, Avoidance of water infrastructure = INR 50 000, Environmental benefits = INR 44 000 Health benefits = INR 793 380 |

[a](base year 2007–2008) USD$1 = Indian Rupee (INR) 42.5.

been necessary to provide water. The cost of digging a borewell, the labour cost and the piping work to connect the borewell to the point of use in the school premises, its maintenance and operation would have cost close to INR 50 000. Apart from this, the third major benefit is towards the environment. Use of treated waste water will reduce the pressure on the surface and groundwater resources, which can be conserved. Treating waste water also prevents environmental pollution and the use of treated waste water for small scale irrigation implies the use of nitrogen and phosphorus that is inherently present in the treated waste water. This leads to less dependency on chemical fertilizers and subsequent cost savings and less environmental pollution. Moreover, the vegetables that were grown using the treated waste water were estimated for cost as per the prevailing market value and the amount of fresh water resource saved because of using treated waste water to grow these vegetables was also considered. All put together, INR 44 000 was the annualized environmental benefits of using treated waste water.

One of the most important and major benefits achieved is health improvement because of the use of safe and clean water. This is due to the reduction in water borne diseases, which are often associated with contaminated water sources. Among a plethora of water borne illnesses, diarrhea is the most common type of health hazard that affects people globally, and according to WHO reports, diarrheal outbreaks in children below 5 years of age are the main cause of mortality in many parts of the world. In this study, the impact of the treated waste water was measured by using two indicators, *i.e.* the reduction in the incidence rates of diarrheal cases per year and by the reduction in mortality rates per year. In this study, it was observed that the number of diarrheal cases could be reduced from 36 to nil in a year by using treated waste water.[12]

The economic benefits of using treated waste water in the school were calculated by the direct expenditure avoided due to less cases of diarrheal disease. This was estimated by using different variables such as the unit cost per treatment, number of cases that were hospitalized, the visits per day per case to seek expert opinion and the hospitalization rate. Moreover, many patients used transport to visit the doctor and therefore there were associated transport costs. Also, the use of treated waste water also means less hospitalization due to improved health, and most importantly, reduced absenteeism in school. Additionally, time saved by the parents or wardens to take care of the sick children and savings in time due to the availability of water facilities in the school (reduction in water collection time) were also taken into consideration for the health benefit analysis that translated monetarily to INR 793 380, as shown in Table 5.7.

#### *5.4.3.4 Conclusions*

Thus, the reuse of treated waste water appears to be very promising as a water conservation strategy given the fact that the natural water resources have been fast depleting in recent years. In this study, it was observed that the benefit of using treated waste water in the school premises far outweighed the cost of the waste water treatment system. Among the various benefit parameters that were studied, the health benefit was the maximum. Thus, savings obtained by avoiding diarrheal diseases in the children and ensuring that they attended school regularly were substantial as compared to the initial cost of constructing the waste water treatment system. Based on these results, it is reported that the stakeholders in the state also sanctioned such treatment systems in other schools as well.

## 5.5 Case Study 5 – Lake Water Treatment in Developing Countries

### 5.5.1 Lake Water Quality in Developing Countries

Surface water is one of the major sources of drinking water in many parts of the world, especially in places where the groundwater table is very low. Lakes are one among the several types of surface water bodies that are usually exploited, and needless to mention, the water quality of these lakes is of paramount importance. However, the lakes in many parts of the globe are not as clean as they should be due to several factors. Some of the main reasons are industrial pollution and dumping of sewage into lakes without appropriate treatment, human activities such as littering and, most commonly, the mixing of domestic waste and sewage into the lake water.

Although there are norms in place in most parts of the world for discharge standards, compliance may also be a key issue, therefore resulting in polluting the lakes. When lake water is the sole source of water for domestic use, its contamination becomes an alarming issue to be tackled. This is even more important if the end use is for drinking, as water consumed can adversely affect human health if it is laden with pollutants such as chemicals and microbes. Therefore, the quality of lake water is a crucial aspect, especially in developing countries where it may be the only source of water and low economic conditions of the people living there may not allow the use of appropriate treatment methods as they may be difficult to afford. The result of this is the manifestation of diseases and an increase in the mortality rates.

UNICEF has pointed out that more than 1.4 million children of age less than 5 years die every year due to diarrhoea, and this is the second largest cause of human death in the world.[23] The situation is even graver in developing countries, as the mortality rate was reported as 1000 per day for a similar age group in 2014[24] and this is only increasing with time. This clearly indicates that the existing methods or technologies are not adequate to treat the waste water discharged in surface bodies like lakes to the approved discharge standards.[25]

Several technologies have been developed to address this problem, however, most of these technologies are effective only at laboratory scale, or they may not be efficient and cost-effective on a large scale. One of the many options that can be used to treat polluted lake water is the use of Hydrodynamic Cavitation.

### 5.5.2 Hydrodynamic Cavitation for Lake Water Treatment

Hydrodynamic cavitation is a green room temperature Advanced Oxidation Process (AOP) that can overcome the present water problem in terms of disinfection, purification, energy efficiency and affordability to satisfy the human need. It is a technology that has been successfully developed at a laboratory scale, and it has also proven to be effective on a pilot scale for various applications, *viz.* extraction, crystallization, degradation and water purification, due to its process intensification effects in terms of mass and heat transfer (transport properties) under atmospheric conditions.[26] However, it still lacks implementation in large/commercial scale applications.

Cavitation is basically a phenomenon of nucleation, subsequent growth, and the implosion of vapour/gas-filled cavities. This can be achieved by alterations in the flow and pressure (hydrodynamic cavitation) or the passage of ultrasound (acoustic cavitation). In hydrodynamic cavitation, the flow geometry is altered by having a flow restriction in such a way that the kinetic energy is increased, which results in a considerable reduction in the local pressure of the liquid. When the local pressure in the constriction region approaches the vapour pressure of the water, millions of vapour cavities are created, which are subjected to turbulent conditions of varying pressure fields downstream of the constriction. The lifespan of the formed cavities is very small (~0.3 microseconds). These cavities finally enter a maximum adiabatic compression phase and result in the generation of very high temperatures (10 000 K) and pressures (up to 1000 bar) inside the cavities, intensifying the chemical reactions or promoting radical formation and its subsequent reactions.[27]

The resultant generated shear forces after the collapse of the cavities coupled with shock waves generated by cavitation events help in breaking down the large pollutant molecules, especially complex large size compounds. The intermediate compounds are more prone to $OH^-$ radical attack and susceptible to easy biological oxidation, which further enhances the overall rate of degradation/mineralization of the pollutant from the waste water. Under such extreme conditions, water molecules within the cavity dissociate into $OH^-$ and $H^-$ radicals. OH radicals are distributed into the water on the complete collapse of the cavities and react with the pollutant molecules, resulting in the mineralization of the pollutants.[28]

It is interesting to note that cavitational methods are reported to be cost effective too. In one such study,[29] Jyoti and Pandit (2001) estimated the cost of the treatment for different types of cavitational reactors compared to the costs associated with conventional disinfection methods. It has been reported that the use of hydrodynamic cavitation using a high-speed homogenizer or orifice plate set-up is a cost-effective treatment strategy (cost of treatment of US\$0.81 and US\$1.4 $m^{-3}$, respectively) as compared to sonochemical reactors or high-pressure homogenizers (cost of treatment of US\$14.88 and US\$6.55 $m^{-3}$, respectively). Moreover, cavitation has also been successfully used for waste water treatment, for instance,[27] Badve *et al.*, in 2015, successfully utilized hydrodynamic cavitation as a promising technology for the treatment of waste water from the wood finishing industry for the removal of volatile organic compounds (VOCs) to reduce its eventual COD.

Thus, cavitational methods can be considered to treat surface water bodies such as lakes that are polluted with various contaminants that arise due to agricultural runoff, human activities and the discharge of industrial effluents. Moreover, cavitation appears to be an affordable technology/process that can disinfect water to avoid water borne diseases. The following section presents a case study where hydrodynamic cavitation has been used to treat polluted lake water.

## 5.5.3 Rankala Lake Clean-up Drive Using Made-in-India Technology – An Initiative by the Swachh Bharat Mission

### *5.5.3.1 About the Swachh Bharat Mission*

The Swachh Bharat Mission or the Clean India Mission is a campaign in India that aims to clean up the country. This mission is run by the Government of India, and its main objective is to eliminate open defecation through the construction of household-owned and community-owned toilets and launching an accountable mechanism of monitoring toilet use. The mission also includes a clean-up drive of all the streets, infrastructure and water bodies of India's cities, towns and rural regions, and it optimistically aims to achieve these objectives by October 2, 2019 (150th birth anniversary of Mahatma Gandhi). It is interesting to note that this mission will also contribute to India reaching the Sustainable Development Goal Number 6 (SDG 6). Launched by the Prime Minister of India, Narendra Modi, the Swachh Bharat Mission is one of the largest cleanliness drives undertaken to date that has several million participants from all walks of life including volunteers, government bodies and NGOs, who are striving to achieve the objectives of this mission. Along similar lines, this case study was undertaken by the author and his team to clean Rankala Lake, thus contributing towards the Swachh Bharat Mission.

### *5.5.3.2 Background*

Rankala Lake, located in Kolhapur, Maharashtra, India has long been a tourist attraction (Figure 5.6). Furthermore, Kolhapur was the capital city of the Maratha Empire during the 16th and 17th century and it is therefore a tourist attraction due to its historical importance. There are 8 quarries that cover the entire Rankala Lake, and it was found that many quarries caused considerable pollution due to the extensive release of untreated sewage and other human activities. For this case study, one quarry having a water holding capacity of about 15 million litres was selected for the treatment. The physical appearance of the water present in the quarry before the

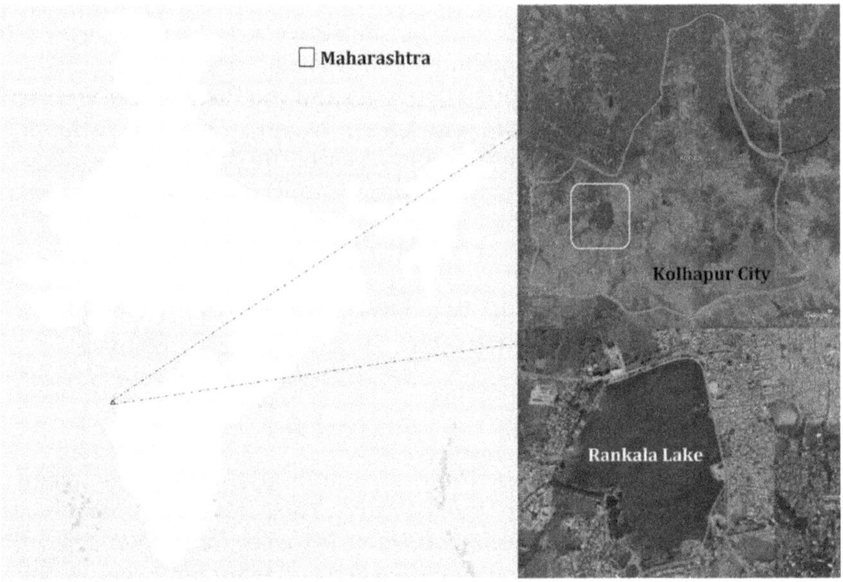


**Figure 5.6** Location of Rankala Lake in India. Map data: Google, DigitalGlobe ©2018 Google.

treatment was greenish black with an obnoxious odour. It also contained a high level of TDS, making this section of the water body quite appalling. Hence, this quarry was selected for the study, and hydrodynamic cavitation was employed to validate this unique technology at large scale.

### *5.5.3.3 Treatment of Rankala Lake*

The hydrodynamic cavitation reactor used for the treatment of Rankala Lake water is schematically shown in Figure 5.7. The design of the cavitating reactor used for this treatment is not revealed here to reserve the right of the patent.[30] The parameters optimized previously by the authors in a pilot scale plant were used for the large scale purification of Rankala Lake water. Specifically, the water from the lake/reservoir was pumped using a centrifugal pump and passed through the cavitating reactor that is shown in Figure 5.7 under specific operating conditions (operating pressure = 4 bar, volumetric flow rate = 100 $m^3 h^{-1}$). The quarry having a capacity of about 15 million litres of water was treated in about 3 weeks (running on an average of 8 h per day). A photograph of the actual equipment used for the study can be seen in Figures 5.8 and 5.9. As described in the previous section, this water purification phenomenon is essentially based on cavitation, in which intensive collapse of the cavity occurs, which results in physical as well as chemical effects resulting in the destruction/reduction of various physical, chemical and biological water contaminants.[31] These effects were monitored in terms of the change (reduction) in the COD, BOD, and the microbial count (CFU per ml) in the treated water.

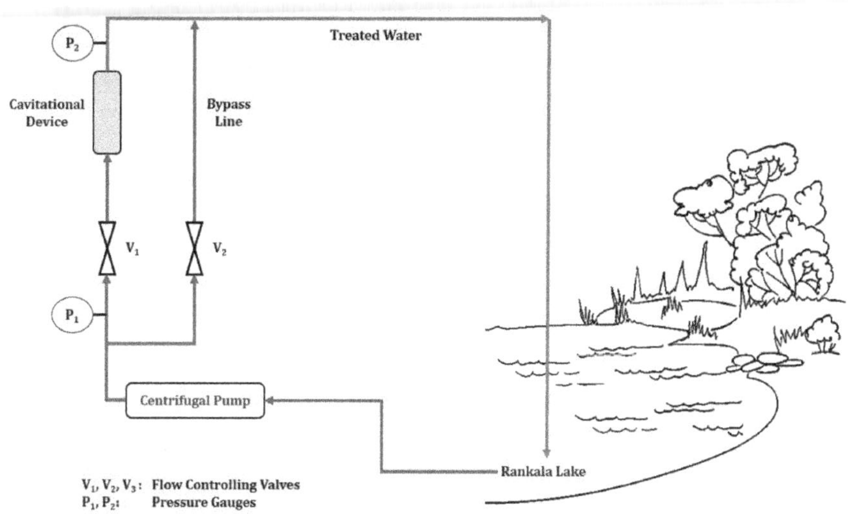


**Figure 5.7** Schematic set-up of hydrodynamic cavitation treatment for Rankala Lake Water.

**Figure 5.8** Hydrodynamic Cavitation System installed at Rankala Lake.

**Figure 5.9** Various Parts of the Hydrodynamic Cavitation System.

#### 5.5.3.4 *Effect of Hydrodynamic Cavitation on Biological Contaminants*

It is very interesting to note that the microbial count of the lake water decreased significantly when treated with hydrodynamic cavitation. The microbial count expressed as CFU per mL was reduced from an initial value of the order of $10^5$ (CFU per mL) to $10^2$ (CFU per mL), indicating a significant 3 log reduction in bacteria (Table 5.8). Additionally, the fungal growth present in the water body was also reduced. This was mainly attributed to the shock waves generated during the intensive collapse of the cavities, resulting in the penetration of these waves through the microorganisms, thereby killing them. Also, the degassing effect, caused by the localized vacuum creation in the cavitator, resulted in total odour removal.

Water purification by hydrodynamic cavitation takes place due to the physical effects generated by cavitation in the form of symmetric as well as asymmetric collapse of the cavities. It is considered that the asymmetric collapse of a cavity is helpful in generating such physical effects to a higher extent. Scientifically, the adiabatic collapse of the cavity should occur only in liquid medium (*i.e.* symmetric collapse of the cavity) to utilize the whole dissipated/released energy during the collapse of the cavities for useful transformations in the liquid medium. However, researchers have observed that if the cavity collapses near a solid surface (*i.e.* the asymmetric collapse of the cavity), then microbial disinfection is of higher magnitude due to the jet cutting action. In this case study, some amount of released energy from the collapsing cavity was lost due to an impact on the solid walls of the cell surface. The wall surfaces may absorb a certain amount of released energy. However, a possible reason might also be due to the continuous turbulence generated in the diverging section of the cavitating device in water because of a series of continuous collapsing cavities that may also de-agglomerate the cell mass of the microorganisms. Hence, the asymmetric collapse may have contributed to a higher disinfection efficacy over the symmetric collapse. Symmetric collapse is also helpful to kill microorganisms that are likely to be present in the bulk of the liquid, possibly due to the oxidation of the cell fluid contents by the generated $OH^-$ radicals.

#### 5.5.3.5 *Effect of Hydrodynamic Cavitation on Chemical Pollutants*

To ascertain the effect of hydrodynamic cavitation on the chemical impurities found in Rankala lake, some essential analytical tests were carried out in the case study, which included COD ($mg\,l^{-1}$), BOD ($mg\,l^{-1}$) and TDS (ppm), turbidity (NTU), chloride content (ppm as Cl) and hardness (ppm as $CaCO_3$) measurement. The effects of cavitation on these parameters are demonstrated in Figure 5.10, which shows the experimental results before

**Table 5.8** Highlights of Rankala Lake clean up case study conducted in Kolhapur, Maharashtra, India.

| Treatment method used | Treatment time | Colour | Odour | Microbial Count | COD | BOD | Power requirement | Cost |
|---|---|---|---|---|---|---|---|---|
| Hydrodynamic cavitation | 3 weeks of treatment (running on an average 8 h per day) | Initial greenish to bluish grey after treatment | Complete odour removal within one week of treatment | Reduced from an initial value of $10^5$ (CFU per mL) to $10^2$ (CFU per mL) | Reduced from 110 ppm to 30 ppm | Reduced from initial 40 ppm to zero | 8.34 kW h | Around 17.52 INR per $m^3$ water and total cost was INR 262 710 |

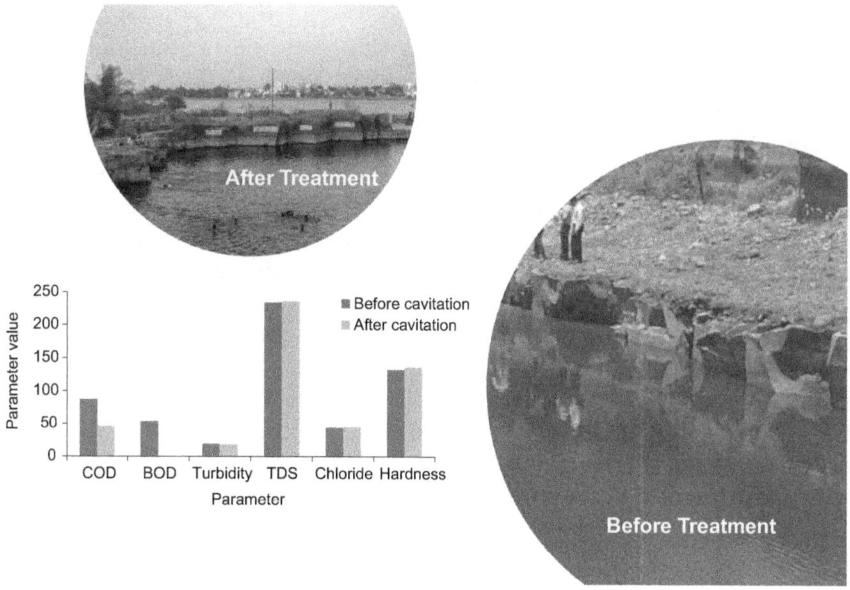


**Figure 5.10** Effect of Hydrodynamic Cavitation on Rankala Lake water characteristics.

and after the treatment, and the highlights of the treatment results are presented in Table 5.8.

The generated OH* radicals due to pyrolysis or dissociation of the water vapour inside the cavities degrade the organic contaminants present in the water body by oxidizing them. Secondary radicals are also generated in the bulk water, which further enhances the degradation of the pollutants. A complete degradation (mineralization) of an organic pollutant gives final products of $CO_2$ and $H_2O$. There might be chances of formation of intermediate products due to partial degradation of those pollutant molecules. However, intermediate products were not analysed as the case study objective was just to convert non-useful water into a useful form.

The initial Chemical Oxygen Demand (COD) of about 110 ppm associated with the lake water was reduced to less than 30 ppm after 6 days of treatment. The Biological Oxygen Demand (BOD) was also reduced from 40 ppm to less than the detectable limit (0 ppm). Intriguingly, the physical appearance of the Rankala Lake quarry water also changed from greenish to bluish grey after the treatment with hydrodynamic cavitation (Figure 5.10). This could also be due to the disruption of algal cell mass (which gave the green colour to the water). The disruption of the algal cells makes the algae sink to the bottom of the lake. This prevents them from receiving sunlight, preventing photosynthesis and algal growth. This amazing change in the appearance and the quality of the lake water encouraged the locals in the city to use the quarry as a swimming pool (Figure 5.10), which is a significant development by itself.

Interestingly, it was observed that the TDS of the water remained the same. The basic mechanism illustrates that only the organic matter gets degraded/reduced due to oxidation, using hydroxyl radicals. This degradation can either occur at the bubble (cavity) interface due to primary radicals generated in the bubble or in bulk by secondary radicals. The metallic compounds cannot be removed by this mechanism, but they can be fragmented into smaller ones, which could further act as nuclei for enhanced bubble (cavity) formation. Because of this, there is an increase in the bubble population in the water, which helps to directly increase active cavitational zones/cavitational density (*i.e.* the number of cavitational events per unit volume of water per unit time) in the water medium.[32] Therefore, some TDS are also indirectly helpful for water purification. Hence, the total concentration of the TDS remains the same, but again, it depends on the nature of the TDS. A similar observation can be made in the case of the chlorides, hardness and suspended solids as well. In addition, chlorides are also beneficial for water disinfection as they act as a disinfectant.

#### 5.5.3.6 *Energy Requirement*

Previous studies by the authors have shown that a 3-bar operating pressure and 30 cycles (circulating all the water through the cavitation device 30 times) are ideal processing conditions for waste water treatment using hydrodynamic cavitation. However, these conditions may change depending on the nature and concentration of the pollutants present in water. Hence, energy and cost estimation in this case study was carried out based on the measured conditions. In the current process, the energy requirement was considerably less as only a centrifugal pump was used, which consumed electrical energy to pass the lake water through the cavitating reactor.

Power utilized by the centrifugal pump for the treatment ($P$) is

$$
\begin{aligned}
P\ (\mathrm{kW}) &= \text{Pressure drop across the cavitating reactor (kPa)} \\
&\quad \times \text{output volumetric flow rate } (\mathrm{m^3\ s^{-1}}) \\
&= \mathrm{d}P \times Q \\
&= (3 \times 10^2) \times (100\ \mathrm{m^3\ h^{-1}}) \\
&= 3 \times 10^2 \times 2.78 \times 10^{-2}\ (\mathrm{N\ m^{-2} \times m^3\ s^{-1}}) \\
&= 8.34\ \mathrm{kW}
\end{aligned}
$$

The processing period was 8 h per day.

$$
\begin{aligned}
\text{Power in kW h} &= 8.34 \times 1\ (\mathrm{h}) \\
&= 8.34\ \mathrm{kW\ h}
\end{aligned}
$$

The electricity cost was assumed to be 7 INR per kW h.

The cost required to pass 100 l of water through the cavitating reactor once

$$= 8.34\ (\text{kWh}) \times 7\ \text{INR per kW h}$$

$$= 58.38\ \text{INR} / 100\ \text{m}^3$$

$$= 0.5838\ \text{INR per m}^3 \text{per pass}$$

The pilot scale experimental results showed that 10 and 30 passes are the ideal processing conditions for efficient treatment for drinking and waste water, respectively, using hydrodynamic cavitation. Beyond these numbers, processing may not be economically feasible. Hence, 30 passes were chosen for the treatment of Rankala Lake water.

$$= 0.05838 \times 30$$

$$= 17.52\ \text{INR per m}^3/30\ \text{passes (nearly US\$0.25 m}^{-3})$$

### 5.5.3.7 *Economics of the Treatment Process*

For any treatment technique to be feasible, it must be cost-effective, and this is an especially important aspect when the treatment is targeted at developing countries. Therefore, the economics of using hydrodynamic cavitation for treating the Rankala Lake water was also studied. The estimated cost of the treatment was around INR 17.52 $m^{-3}$ ($\approx$25 cents $m^{-3}$) water, and this is far more reasonable than other cavitational[29] as well as conventional waste water processing techniques.

For 15 million l of water, the cost

$$= 17.52 \times 15\,000$$

$$= 262\,710\ \text{INR}$$

$$= 3725\ \text{USD (\$)}$$

Thus, the total estimated treatment cost of that Rankala Lake quarry was about INR 262 710 or (3725 USD).

### 5.5.3.8 *Conclusion*

Thus, effective treatment of surface water bodies like lakes can result in the preservation of nature's gift and ensure optimal use of natural resources. This was sufficiently demonstrated in the case study where Rankala Lake (Quarry) having 15 million litre water capacity was successfully and efficiently treated using hydrodynamic cavitation. A significant change in the physical appearance of the water from green to bluish grey was observed, and the foul smell emanating from the water was also removed completely after one week of the treatment. Moreover, the treatment significantly reduced bacteria, colour as well as organic contaminants present in the lake water. Currently, the people living around the lake are using the quarry as a swimming pool, which is strong proof of the effective treatment. In scientific terms, the bacterial reduction was achieved from $10^5$ CFU per mL to $10^2$ CFU per mL, and this almost 3 log reduction satisfies the expectations of the Central Pollution Control Board. A major reduction was observed in BOD (100%), COD (72%), and microbial population (99.9%) with respect to their initial values. Energy and cost estimation (17.52 INR per $m^3$) show that hydrodynamic cavitation is an emerging technique in the technological development of waste water treatment on a larger scale.

## 5.6 Case Study 6 – Disinfection of Borewell Water Using Hydrodynamic Cavitation in India, the Mark II Hand Pump

### 5.6.1 Ground Water Quality in Developing Countries

The previous case study discussed lake water treatment and like surface water bodies, groundwater is also a major source of water in many parts of the world. Groundwater is the principal source of potable water and it is an indispensable source of life. Groundwater is formed due to the seeping of rain water down through the different layers of the earth. Rivers, lakes and surface water bodies also contribute to groundwater. To access the groundwater, usually a bore is drilled underground and hence the name bore water is frequently used synonymously with groundwater. Groundwater is used for several domestic purposes like cooking, cleaning, and personal use like bathing, recreation, *etc.* Most commonly, groundwater is also used for drinking and therefore its quality is of utmost importance. Depending on the geographical location and the environmental conditions, the overall ground water characteristics may vary, thereby affecting its suitability for various uses. Groundwater may naturally contain certain minerals, chemicals and microorganisms that may be harmful if consumed without appropriate treatment. However, human or other anthropological activities are also known to pollute groundwater and render it inappropriate for drinking.

Contamination of groundwater may occur in several ways. For example, agricultural activities that use fertilizers, pesticides *etc.*, may result in seeping of these chemicals underground. Other waste disposal systems and septic tanks that are not managed effectively can also contribute to pollution of groundwater. Contamination due to mixing of sewage with borewell water is a major concern in many parts of the world and is more pronounced in developing countries because of a lack of proper sewage treatment systems and inadequate maintenance of the borewells, which are often the sole means of potable water for the people inhabiting these rural areas. Most often, a contaminated borewell may go unnoticed as the water may not show visible signs of pollution. However, an unusual taste, odour or colouration of the borewell water may be a sign to be watched for by the people living there. An immediate quality check carried out by concerned water authorities can throw light on the physical, chemical and biological quality of the water. Thus, constant monitoring is a necessity to ensure the safety of the people consuming water from such borewells and this may not always be feasible and in many cases, the quality may be poor, requiring some type of treatment before it is used by the inhabitants.

Simple, effective and affordable treatment technologies that can be easily applied to borewells in developing countries are always a pressing need. Among the various treatment methods that are routinely used and have been extensively discussed in this book, hydrodynamic cavitation appears to be promising. In the previous case study, it was clearly shown that hydrodynamic cavitation could effectively treat polluted lake water.

The hand pump can serve as an efficient tool for the disinfection of borewell water, with specific modification in the geometry of the check valve. This modification leads to the phenomenon known as hydrodynamic cavitation, which can potentially disrupt bacterial cells. This mechanism is more energy efficient compared to homogenization and acoustic cavitation. Hydrodynamic cavitation technology for the hand pump has recently emerged for application in water disinfection. This novel technology was validated by conducting a pilot study as well as field trials and was found to have yielded more than 95% disinfection. By using these modified hand pumps, people from rural areas across the world will have access to good quality potable water and can be saved from a number of water borne diseases. Hand pump cavitation will be a milestone in the emerging era of novel and economical water disinfection systems. Therefore, in this case study, hydrodynamic cavitation that occurs in hand pumps has been investigated for borewell water treatment.

### 5.6.2 Why the India Mark II Hand Pump?

In the previous case study on lake water treatment by hydrodynamic cavitation, a brief discussion on the principle and concepts of hydrodynamic cavitation has already been included. However, for the benefit of readers who are specifically going through the present case study, a small description of the phenomenon is explained here. It is interesting to note that cavitation was first discovered by Reynolds in the year 1873 due to the unusual behavior of boat propellers at high rotational speeds. In the year 1906, Parsons improved upon Reynolds concept, recognized the role played by liquid vaporization in cavitation, and conducted the first experiments on cavitation. Cavitation is a stochastic phenomenon and it is defined as the inception, growth and adiabatic collapse of a partially or completely condensable gaseous or vapour cavity in a liquid. The collapse of cavities results in very high temperatures and pressures, locally rising to 12 000 K and 5000 atm, respectively. Cavitation occurs when the pressure of the liquid falls below the vapour pressure of the liquid being subjected to flow.

Several studies have reported that hydrodynamic cavitation can be effectively used to kill microbial cells. Use of hydrodynamic cavitation for cell disruption was reported by Harrison and Pandit (1992),[33] Save *et al.* (1994 and 1997),[34,35] Balasundaram and Pandit (2001),[36] Sivakumar and Pandit (2002)[37] and Manchalwar (2015),[38] In the previous case study on Rankala lake water treatment, use of hydrodynamic cavitation successfully reduced the microbial count. Moreover, several studies conducted by the authors as mentioned in the previous case study have shown that hydrodynamic cavitation is an effective way to inactivate microbial cells and can used in water disinfection as well. Hand pumps are widely used in rural areas of developing countries to draw ground water. The India Mark II Hand Pump (Figures 5.11, 5.12 and 5.13) is suitable for lifting water from depths of up to 50 m and can be installed in borewells as well as dug wells. The India Mark II Hand Pump was developed in India more than 30 years ago. In Indian villages, hand pumps are widely used and sometimes they are used continuously for up to 18 hours a day.

The author has initiated a major research programme on cavitationally induced transformations and developed the novel technique of hydrodynamic cavitation. He is involved in the development of hydrodynamic cavitation prototypes that are undergoing field trials. In a pioneering attempt, modifications of the hand pump to create cavitating conditions for borewell water disinfection were carried out by the author and his team.

### 5.6.2.1 *Working Mechanism of the Hand Pump*

The mechanism of the hand pump is based on the working principle of a positive displacement pump, negative pressure being the potential behind lifting of the water from the well to the surface level. The cylinder assembly of a hand pump consists of a cast iron cylinder body, with two caps, and an operating mechanism like a plunger yoke body with a follower, spacer-upper check valve assembly along with a check valve assembly that moves up inside the cylinder. At this point, the lower check valve is closed, thus allowing the water from the well to flow inside the cylinder. On the return stroke when the handle is moved up, the plunger assembly starts coming down inside the cylinder. The check valve at the bottom will close, thereby forcing the water through the follower upwards opening the upper valve. Thus, when the plunger moves up and down in the cylinder, the water is displaced and finds its way to the overhead water tank assembly.

### 5.6.2.2 *Cavitation for Microbial Disinfection*

Cavitation is created through a restriction in the flow area opening as a function of lift of the check valve. When cavities collapse, they let out a high

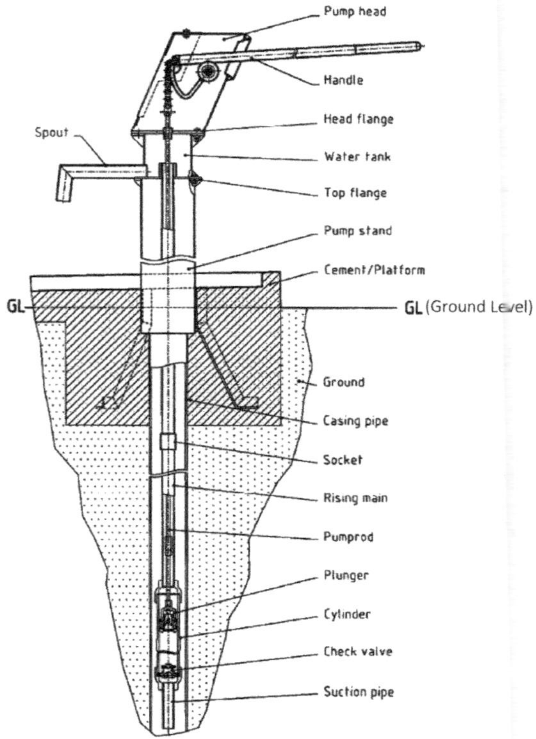


India Mark II Hand Pump Head Assembly

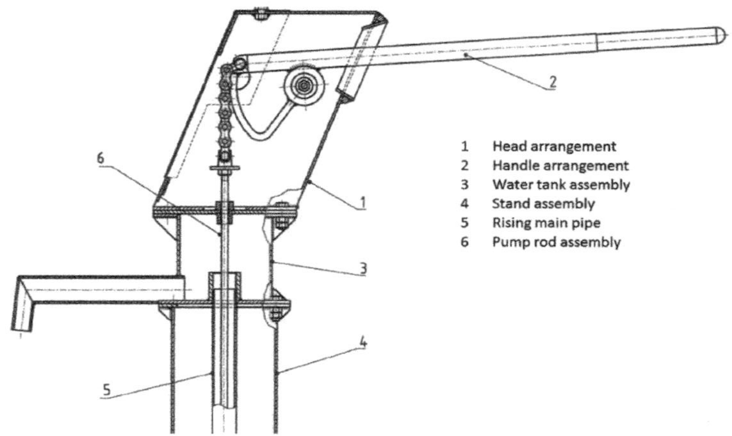


India Mark II Hand Pump Head Assembly

**Figure 5.11** India Mark II Hand Pump installed in a borewell.

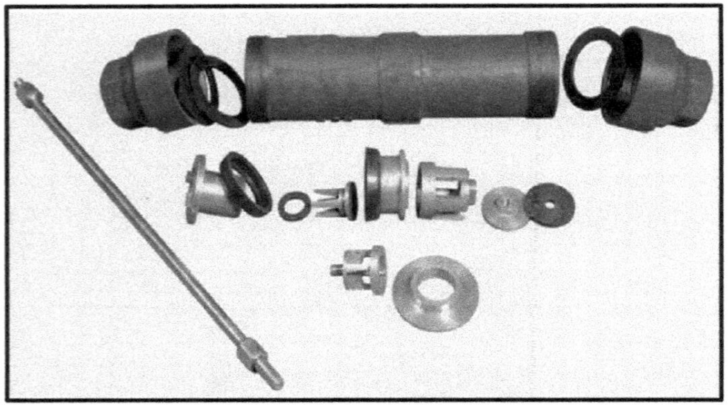

Photograph of India Mark II Hand Pump cylinder internals

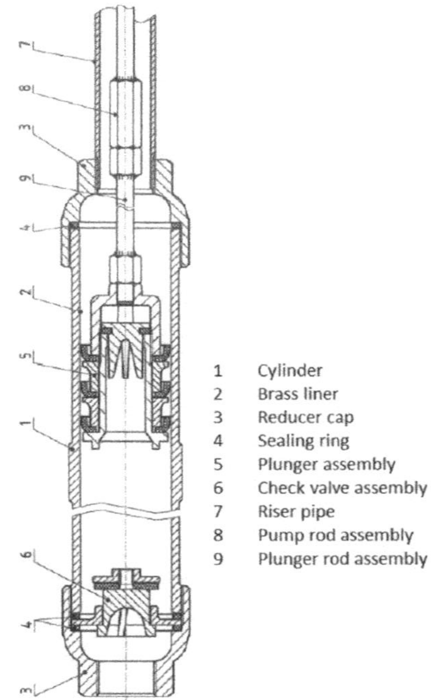


Schematic representation of India Mark II Hand Pump cylinder internals

**Figure 5.12** India Mark II Pump cylinder.

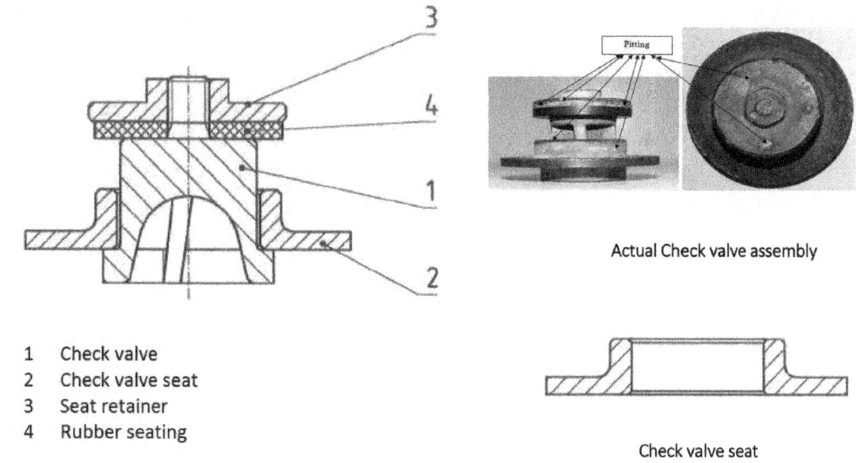


**Figure 5.13** Original check valve assembly.

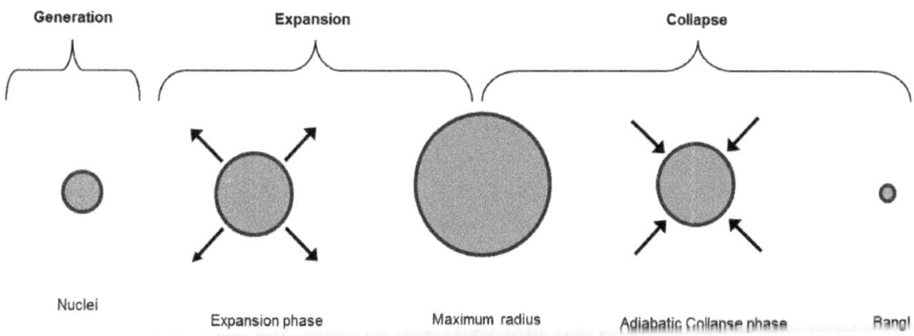


**Figure 5.14** Mechanism of cavitation.

velocity microjet and a spherical shockwave that, if the former is incident on a cell wall of a microbe, will cause the cell wall to disrupt and the microbe will be rendered inviable (Figure 5.14). Intensity of cavitation was optimized in the hand pump to give a maximum disinfection in a single pass. The flow area opening as a function of lift was restricted by modifying the check valve. The generation of vapour cavities as well as the size of the cavity increases during the upward stroke of the hand pump due to the pressure fall (*i.e.* negative pressure). All the vapour cavities will collapse simultaneously due to the huge back pressure generated during the downward stroke. Many of the cavities will collapse automatically due to overgrowth; still, the population of cavities will be high. Probably, asymmetric collapse is more possible by reason of the higher lifespan of the cavities. Thus, mechanical as well as chemical effects are high in the hand pump cavitation. This is a positive outcome of a discrete back pressure system (*i.e.* hand pump) rather than a continuous one (*i.e.* existing hydrodynamic cavitation system).

While in a continuous back pressure system, the cavity size will be smaller and the lifespan will be shorter. The symmetric collapse is more possible due to an earlier collapse. Hence, it may have limited effects as well as the action of the cavity being limited in existing/conventional cavitational devices (*i.e.* nozzle, venturi, orifice *etc.*). The actual disintegration of a microorganism occurs due to two mechanisms: (1) intensive collapse of the cavity (spherical shock wave) and (2) the high velocity of microjets generated during the strong asymmetric collapse of the cavities {the dominant form of perturbation (Rayleigh–Taylor instability) is a re-entrant jet}. If the re-entrant jet or the shockwave is incident to the microbial cell membrane present in continuum, this small injury disintegrates the cell membrane and renders the microbe (bacteria, fungi, viruses) unviable. There are various cavitational devices developed to treat water. The use of hydrodynamic cavitation in hand pumps is a new practice that is economically inexpensive and extensively studied.

Generation of back pressure is the key factor for disinfection of the bacteria, viruses and fungi. But back pressure should not affect the cavities during their growth because the cavities will not get an opportunity to grow adequately, resulting in smaller size cavities that require more back pressure to burst intensively as well as their effect being limited (*e.g.* a small balloon requires more press from opposite sides to burst than a large one). As the back pressure increases, the expansion slows down. The minimum expansion of a bubble should be more than 2 times the initial size for cavitation to occur. Earlier collapse of the bubble can occur due to back pressure, hence the collapse probability of the bubble is more in the bulk rather than on the cylinder wall surface. Therefore, symmetric collapse will often be more in the case of a higher back pressure. Whereas, low pressure introduces higher expansion as well as increasing the lifespan of the bubble. Hence, a bubble can travel up to the solid wall surface and collapse, which will be asymmetric in nature. Mass transfer rates are zero for pressures greater than 3 atm, thus only mechanical effects are possible in the case of an ultrasound system. Hence, a lower pressure is preferable to enhance the chemical effects. The hand pump cavitation is balanced for physical and chemical effects due to the large pressure drop generated during the upward and downward stroke.

For a higher opening of the valve, the probability of the generation of vapour cavities is lower due to the low velocity or high partial pressure upon opening the valve. There may be the generation of only gaseous cavities, which are limited in action as well as quantity. In addition, hydrostatic pressure could be higher, which affects the cavitational growth, and consequently there will do less disinfection. While for a lower opening of the valve, cavitational growth is more due to additional water evaporation.

This may have four consequences, enlisted as below.

(1) A buoyancy force acts on the overgrowing cavities, which pulls them towards the upward surface of water and nullifies the effect.

(2) There may be chances of the formation of bubble clusters/clouds because as the size of the bubbles increases, interaction between them increases and they form clusters, which can dampen the cavitation.
(3) The cavities may burst automatically in the bulk before the downward stroke (*i.e.* before the generation of back pressure) of a piston or plunger yoke due to their overgrowth, resulting in a less intensive collapse that is not enough to kill the bacteria.
(4) The very high velocities will decrease the number of cavities upon the opening of the lower check valve.

Finally, cavities should grow sufficiently large and collapse simultaneously after the action of back pressure applied by the piston during the downward stroke. Simultaneous collapse generates highly intensive shock waves as well as micro jets during asymmetric collapse resulting in more disinfection.

## 5.6.3 Experimental Investigation

### *5.6.3.1 Background*

The aim of this research activity was to deliver disinfected water (or water with pathogens within the prescribed limits or to reduce the number of pathogens that are responsible for causing water borne diseases) by a modified hand pump design. To achieve this, it is necessary to understand the underlying principles of pumping, cavitation, and disinfection due to cavitation, and the mechanism and hydrodynamics of the hand pump. The following studies have been carried out:

1. Standardization of CFU counting method.
2. Design of the India Mark II Hand Pump was studied in detail.
3. After extensive numerical simulations, it was predicted that specific modification generates the maximum cavitational effect that causes maximum pathogen destruction.
4. Different modifications were fabricated and tested first using the pilot set-up and then in the field in rural locations.

Computational fluid dynamics simulations were used to optimize the dimensions of the modification. ANSYS 16.0 FLUENT was used to simulate the geometry of the India Mark II hand pump. To optimize the dimensions of the check valve, the percentage flow area opening as a function of lift that produced the maximum intensity of cavitation was investigated. Experiments were carried out to validate the results using a pilot set-up installed at the Institute of Chemical Technology (ICT), Mumbai, Maharashtra, India as well as in other rural villages where hand pumps are used to draw water from borewells. The hand pump used in this case study was the "India Mark II hand pump". This hand pump was modified based on CFD analysis for creating cavitating conditions and used to disinfect borewell water that passed through it.

#### 5.6.3.2 Pilot (Replica) System in the ICT Premises

Based on the typical water pumping rates obtained using an India Mark II hand pump operating at 50 strokes per minute, inlet velocity was found to be 3.1 m s$^{-1}$. From the specifications provided of the India Mark II hand pump, hydraulic diameter was found to be 0.04 m. Samples were collected from the pilot plant installed on the ICT premises (Figure 5.15) with pumping rates of 40 strokes, 50 strokes and 60 strokes per minute and independently stored. A sample bottle was filled with the original spiked water before passing the same through the pump. There was a certain concentration of *E. coli* bacteria in the original sample. Then, water was drawn from the pump and the first 10 liters of water was discarded. This was done because water stagnancy would have caused the growth of microbes within the pump cylinder. The handle of the hand pump was moved up and down 40 times during a minute. The water coming out of the pipe of the hand pump was collected as the "40 strokes" sample. The procedure was repeated for 50 strokes and 60 strokes per minute. CFU analysis of the collected samples was then carried out as per the standard protocol. A water sample was spiked with a fresh culture of *E. coli* and the sample was collected and tested before and after treatment to check the repeatability of efficacy of the treatment. Essentially, the treatment consisted of pumping the spiked water using different India Mark II modified hand pumps. The culture used for spiking was a 24 h fresh culture of *E. coli* (ATCC 2346) with a culture density of $10^9$ CFU per ml. 100 ml of this culture was inoculated into 100 L of raw water in a tank to get the final concentration of approximately $10^6$ CFU per ml. Enough care was taken to ensure that the culture was mixed properly in the raw water.

Spiked raw water (before treatment sample) was collected in a sterile bottle and brought to the laboratory in an ice pack cooler bag at 0 °C to 4 °C. Subsequently, serial dilutions were carried out and each dilution was

**Figure 5.15** Set-up site on ICT premises.

inoculated in sterile petri dishes poured with sterile Plate Count Agar, and the plates were incubated at 37 °C for 24 to 48 h. After completion of incubation, the colonies were counted in the raw and treated water samples. The following formula was used to calculate % reduction in bacteria:

$$\text{Percentage Reduction} = \frac{\text{Count in raw water sample} - \text{Count in treated water sample}}{\text{Count in raw water sample}} \times 100$$

Tables 5.9–5.11 show the results obtained in the pilot trials carried out on the ICT premises. It can be observed that the percentage reduction of microorganisms was higher when modified check valves were used in the hand pump and almost 90% inactivation could be achieved. Interestingly, the number of strokes of the hand pump also played a vital role in microbial disinfection. With 60 strokes per minute, the stroke efficiency increased from 23.73% (original check valve) to almost 80% for the modified check valve. These promising results led to the conception of field trials where these experiments could be validated further.

#### 5.6.3.3 *Treatment of Borewell Water (Field Trials)*

The experimental location was selected from rural areas where people are prone to water borne diseases due to unavailability of drinking water treatment facilities (Figure 5.16). Initially, samples were taken for the existing system. The hand pump was then disassembled completely, a modified valve was fixed into it and the hand pump was reassembled again. In the beginning, a few strokes were done to remove rust that got mixed in the borewell water during assembling. The water was sampled after the removal of rust by visual inspection (Figure 5.17). Similarly, other modified valves were tested. All samples were stored in sampling bags, which were maintained at a relatively low temperature by ice packs (to prevent any further growth of the bacteria) and were internally coated with thin Teflon sheets to avoid heat loss. The CFU per ml analysis of all samples was done in a laboratory on the ICT premises within 24–48 h.

The following are a few observations made during the field work.

1. Temperature at the field location varied from 28 °C to 36 °C.
2. The capacity of a single pipe of the hand pump assembly is 3 l.
3. The water bucket capacity (avg.) is 13 l (to calculate the water discharge quantity).

### 5.6.4 Sample CFU per ml Analysis

The plate count method means diluting bacteria with a diluent solution (*e.g.* sterile water) until the bacteria are dilute enough to count accurately

**Table 5.9** Effect of modified check valve assembly on the CFU per ml behaviour of the borewell water.

| Samples | No. of strokes | Dilutions 10 | 100 | 1000 | 10 000 | CFU per ml | Percentage disinfection | Average disinfection |
|---|---|---|---|---|---|---|---|---|
| Original | — | C* | C* | 113 | 2 | 113 000 | * | — |
| With modified check valve | 40 | C* | 118 | 25 | 1 | 11 800 | 89.56 | 89.97 |
| | | C* | 112 | 18 | 0 | 11 200 | 90.09 | |
| | | C* | 110 | 21 | 4 | 11 000 | 90.27 | |
| | 50 | C* | 101 | 19 | 0 | 10 100 | 91.06 | 91.06 |
| | | C* | 103 | 16 | 1 | 10 300 | 90.88 | |
| | | C* | 99 | 13 | 3 | 9900 | 91.24 | |
| | 60 | C* | 75 | 26 | 2 | 7500 | 93.36 | 91.24 |
| | | C* | 131 | 12 | 5 | 13 100 | 88.41 | |
| | | C* | 91 | 26 | 9 | 9100 | 91.95 | |

**Table 5.10** A typical plate count analysis.

| Sample | Dilutions 10 | 100 | 1000 | 10 000 | 100 000 | 1 000 000 | 10 000 000 | 100 000 000 | CFU per ml | Percentage reduction |
|---|---|---|---|---|---|---|---|---|---|---|
| Original | C* | C* | 0 | 1 | 15 | 0 | C* | 0 | 320 000 | 0 |
| | 82 | 13 | 0 | 32 | 0 | C* | 0 | 0 | | |
| | C* | 4 | 1 | 17 | C* | 0 | 0 | 0 | | |
| 40 strokes | 46 | C* | 6 | 1 | 1 | 0 | 0 | 0 | 3333.33 | 98.96 |
| | C* | 4 | 1 | 0 | C* | 1 | 0 | 0 | | |
| | 55 | 4 | 26 | 0 | 0 | 0 | 0 | 0 | | |
| 50 strokes | C* | C* | C* | C* | 0 | C* | 0 | 0 | 10 000 | 96.87 |
| | 58 | 3 | 0 | C* | 0 | 2 | 0 | 0 | | |
| | 45 | C* | C* | 1 | 0 | C* | C* | 0 | | |
| 60 strokes | 193 | 6 | 0 | 0 | 6 | C* | 0 | 0 | 20 000 | 93.75 |
| | 75 | C* | 1 | 4 | 1 | 0 | 0 | 0 | | |
| | C* | 20 | 6 | C* | C* | 4 | C* | 0 | | |

**Table 5.11** Study of stroke efficiency.

| System | Strokes per min | Volume of water pumped (l) | Theoretical output (l) | Stroke efficiency |
|---|---|---|---|---|
| Original | 40 | 11.140 | 15.82658 | 70.38% |
| | 50 | 13.680 | 19.78323 | 69.14% |
| | 60 | 17.600 | 23.73987 | 23.73% |
| Modification | 40 | 9.096 | 15.82658 | 57.47% |
| | 50 | 13.000 | 19.78323 | 65.71% |
| | 60 | 19.160 | 23.73987 | 80.70% |

**Figure 5.16** Field trials for India Mark II Hand Pump.

Water received immediately after installation
(muddy colour due to complete dismantling and assembling of the pipe below ground level)

Water received after some time

**Figure 5.17** Collecting water after installation of modified valve with calibrated strokes.

when spread on a plate. The assumption is that each viable bacterial cell will develop into a single colony. Bacterial cell numbers need to be reduced by dilution, to be able to count them accurately. To avoid contamination, the sample bottles were stored at 4 °C in refrigerators.

### 5.6.4.1 Material

0.9% (w/v) sodium chloride (HIMEDIA) solution as the diluent, 20 ml test tubes, cotton, petri dishes and the spreader were used. An autoclave, Plate Count Agar (HIMEDIA), 70% ethanol solution, a Laminar Cabinet, a Variable volume micro-pipette along with micro-tips and the Incubator were used. A pair of Bunsen burners in a laminar cabinet (for a sterile environment), thermometer, pH meter, buffer capsules and refrigerator were used. Check valve assemblies with different geometries were fabricated in-house.

### 5.6.4.2 Procedure

The pH meter was calibrated before use with the help of pH buffer capsules, and the pH of each sample was measured. The viable cell count (spread plate) method was used to calculate the Colony Forming Unit (CFU) per ml of the borewell water. Sterilization of all equipment in the autoclave at 121 °C, 15 bars over 15 min was done. Plate Count Agar was used as the medium to grow the bacteria. A laminar flow chamber was washed with 70% ethanol and UV light was switched on for 15 min for laminar cabinet sterilization. All the samples were kept in the laminar cabinet after sterilization to avoid any possible contamination. The variable volume micro-pipette along with micro-tips was used for the serial dilutions. Triplet serial dilutions of each sample were made in between 2 flamed burners with sodium chloride solution. Aliquots (0.1ml) were spread, using a sterile bacterial spreader, onto the agar plates. The agar plates with sample spread over it were incubated for 24–48 h at 37 °C. The number of bacterial colonies that appeared on each of these plates was counted. Also, one blank petri-plate filled with agar (media) was also kept during the analysis to account for any internal contamination during CFU analysis and it showed no colony growth (no CFU).

To compute the estimated total plate count on the solidified agar plate, we used the following formula:

$$\text{Total plate count} = N \times d$$

where $N$ = number of colonies counted on the plate and $d$ = dilution factor.

$$\text{Example}: N = 56 \text{ colonies}; \ d = 1000;$$

$$\text{Total plate count} = 56\,000 \text{ CFU in } 0.1 \text{ ml}$$

$$= 560\,000 \text{ bacteria per ml}$$

From the growth rate of bacteria and incubation temperature, we can predict that the bacteria are mesophilic (Figure 5.18). The higher growth of bacteria occurs at 37 °C, which is related to the mesophilic category while the growth of thermophilic bacteria occurs above 45 °C, and psychrophilic bacteria grow below 30 °C. The temperature of the experimental/trial places supports the prediction.

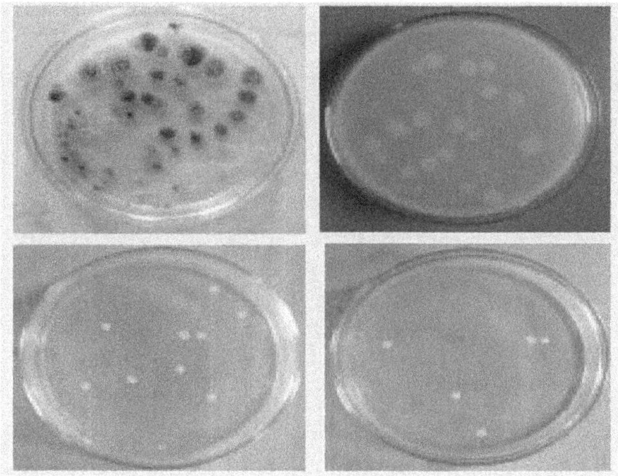


**Figure 5.18** CFU per ml Count observed before (A) and after (B) treatment.

### 5.6.4.3 *Results from EnviroCare Lab, Mumbai, India*

Samples were sent to a Government approved laboratory, EnviroCare Lab, for analysis and calculations were made as follows:

Calculation:

$$\% \text{ Reduction} = \frac{\text{Count in raw water sample} - \text{Count in treated water sample}}{\text{Count in raw water sample}} \times 100$$

$$= \frac{(3.6 \times 10^9) - (4.4 \times 10^6)}{(3.6 \times 10^9)} \times 100$$

$$= 99.87\%$$

Thus, it was concluded that CFU per ml reduction was in the range of 99.87% (Table 5.12).

### 5.6.4.4 *Results from Microbiology Laboratory, Nashik, India*

After obtaining a significant reduction in CFU per ml, it was also necessary to ascertain the type of bacteria present in the water being treated. Therefore, samples were sent to the Microbiology Lab and tests were conducted to determine the type of bacteria.

From Table 5.13, the 6th bacteria (ICT/15/5B) came under the Bacillus category having 3 subtypes. To confirm the exact category, an additional (Inulin) test was carried out.

**Table 5.12** Total bacterial count in raw and treated water samples.

| S. no. | Sample | Incubation condition | Dilutions | Set I (CFU per ml) | Set II (CFU per ml) |
|---|---|---|---|---|---|
| 1. | Raw water {before treatment} | 37 ± 1 °C | Undiluted | TNTC[a] | TNTC[a] |
| | | | $10^{-1}$ | TNTC | TNTC |
| | | | $10^{-2}$ | TNTC | TNTC |
| | | | $10^{-3}$ | TNTC | TNTC |
| | | | $10^{-4}$ | TNTC | TNTC |
| | | | $10^{-5}$ | TNTC | TNTC |
| | | | $10^{-6}$ | TNTC | TNTC |
| | | | $10^{-7}$ | TNTC | TNTC |
| | | | $10^{-8}$ | 350 | 370 |
| | Total count | | | $3.6\times10^{9}$ CFU per ml | |
| 2. | Water after passing through modified India Mark II Hand Pump {after treatment} | 37 ± 1 °C | Undiluted | TNTC | TNTC |
| | | | $10^{-1}$ | TNTC | TNTC |
| | | | $10^{-2}$ | TNTC | TNTC |
| | | | $10^{-3}$ | TNTC | TNTC |
| | | | $10^{-4}$ | TNTC | TNTC |
| | | | $10^{-5}$ | 41 | 48 |
| | Total count | | | $4.4\times10^{6}$ CFU per ml | |

[a]TNTC = Too Numerous to Count.

**Table 5.13** The bacterial identification of treated water.

| S. no. | Sample ID | Bio number | Analysis time (h) | Probability | Bacterial analysis |
|---|---|---|---|---|---|
| 1. | ICT/15/3B | 014010302000000 | 8.25 | 93% | Kocuriarhizophilia |
| 2. | ICT/15/3A | 014010302000000 | 8.00 | 93% | Kocuriarhizophilia |
| 3. | ICT/15/5A | 000010100000000 | 6.00 | 99% | Kocuriarosea |
| 4. | ICT/15/5C | 0252511565453220 | 14.25 | 97% | Bacillus megaterium |
| 5. | ICT/15/5D | 041032310000000 | 5.00 | 98% | Micrococcus luteus |
| 6. | ICT/15/5B | 1070171515647220 | 14.25 | Bacillus subtilis[a]/ amyloliquefaciens/atrophaeus | |

[a]The 6th bacteria (ICT/15/5B) came under Bacillus category having 3 subtypes. To confirm the exact category, additional (Inulin) tests were done (see Section 5.22.4).

Details of 6th bacterial test are as follows:

1. Bacillus subtilis: INULIN (11), Cell Chains (84)
2. Bacillus amyloliquefaciens: Cell Chains (20)
3. Bacillus atrophaeus: INULIN (83), Cell Chains (23)

Thus, Bacterial Identification Automated Antibiotic susceptibility testing (VITEK-2) clearly showed that the opportunistic pathogens such as Kocuria, Bacillus, Micrococcus, *etc.* were successfully eliminated by cavitation effects in the India Mark II hand pump.

### 5.6.4.5 *Results from the Field Trials*

The calculations clearly indicate that 40–60% flow area opening as a function of lift is the optimum dimension of the check valve. This can generate maximum cavitational shear to kill microbes present in water. Thus, the experimental results clearly show that maximum disinfection can be obtained at 40–60% flow area opening, and the numerical simulations indicated maximum mean cavity collapse pressure also at 40–60% flow area opening (Tables 5.14, 5.15, and Figure 5.19). Also, the modification is

**Table 5.14** Different rural locations where trials were conducted.

| S. no. | Study (in collaboration with) | Location (depth) | Application |
|---|---|---|---|
| 1. | ***Ground water treatment***: Local Grampanchyat with ICT Mumbai | A/P Khudus, District Solapur, Maharashtra (80 ft)<br>A/P Atpadi, District Sangli, Maharashtra (100 ft)<br>A/P Bhadas, District Pune, Maharashtra (120 ft)<br>A/P Malshiras, District Solapur, Maharashtra (80 ft)<br>A/P Zanjewaadi, District Solapur, Maharashtra (100 ft) | Hydrodynamic Cavitation for microbial disinfection |
| 2. | **Borewell water treatment** | ICT Mumbai Premises (Pilot study) | Hydrodynamic Cavitation for microbial disinfection |
| 3. | **Borewell water treatment** | Nahur, Mumbai | Hydrodynamic Cavitation for microbial disinfection |

**Table 5.15** Variation of different parameters observed with different modifications.

| System | Strokes per min | CFU per ml | Percentage disinfection | V actual ($m\ s^{-1}$) | Percentage stroke efficiency |
|---|---|---|---|---|---|
| M1 | 40 | 450 000 | 0 (Ref.) | 0.31 | 86.45 |
| | 50 | 420 000 | 6.67 | 0.45 | 99.63 |
| | 60 | 380 000 | 15.56 | 0.53 | 97.16 |
| M2 | 40 | 300 000 | 33.34 | 0.62 | 99.37 |
| | 50 | 20 000 | 95.56 | 0.76 | 96.98 |
| | 60 | 27 000 | 94 | 0.92 | 97.16 |
| M3 | 40 | 80 000 | 82.23 | 0.71 | 96.06 |
| | 50 | 340 000 | 24.45 | 0.86 | 93.27 |
| | 60 | 135 000 | 70 | 1.04 | 94.29 |
| M4 | 40 | 180 000 | 60 | 0.87 | 98.04 |
| | 50 | 310 000 | 31.12 | 1.1 | 99.1 |
| | 60 | 18 000 | 96 | 1.22 | 91.42 |
| M5 | 40 | 43 000 | 90.45 | 1.03 | 92.08 |
| | 50 | 24 000 | 94.67 | 1.23 | 87.71 |
| | 60 | 31 000 | 93.12 | 1.45 | 86.34 |

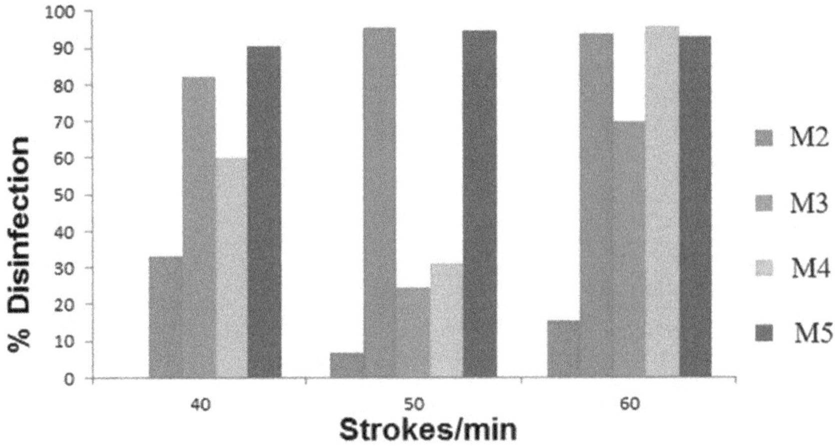


**Figure 5.19** No. of stokes *vs.* percent disinfection.

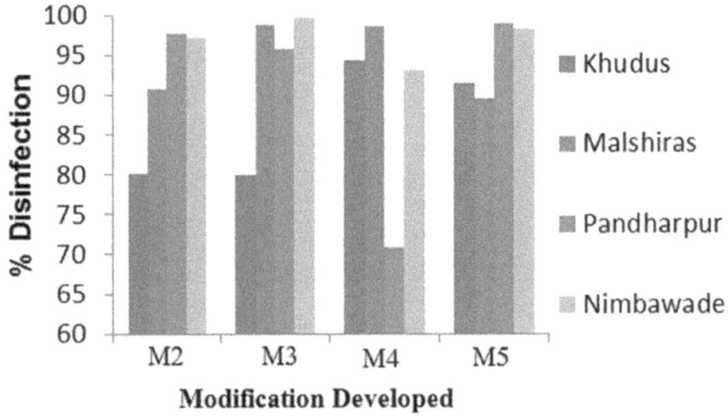


**Figure 5.20** Percentage disinfection observed at different rural locations.

cheaper to implement and easier to manufacture. However, more field trials need to be carried out to assess the locational effect of water quality.

Thus, it was found that the computational fluid dynamics simulations were accurate in predicting the dimensions of the modified check that generated maximum cavitational effects and thus maximum disinfection. For a single stage assembly, 85–99% disinfection is highly desirable (Figure 5.20, Table 5.16). It has also been reported that the strength and structural integrity of pathogenic cell walls are lower than that of *E. coli* bacteria. Thus, 85–90% disinfection would therefore allow for a much higher level of pathogenic disinfection that would result in fewer cases of water borne diseases in Indian villages. It has also been proved that the cavitational shear created by the check valve modification is effective in microbial disinfection.

**Table 5.16** Effect of different modifications on percentage disinfection.

| Trial no. | Location of field trial | System | % Disinfection |
|---|---|---|---|
| 1 | Khudus, Dist.-Solapur | M2 | 80.08 |
| | | M3 | 79.96 |
| | | M4 | 94.39 |
| | | M5 | 91.49 |
| 2 | Malshiras, Dist.-Solapur | M2 | 90.91 |
| | | M3 | 98.78 |
| | | M4 | 98.63 |
| | | M5 | 89.54 |
| 3 | Pandharpur, Dist.-Solapur | M2 | 97.73 |
| | | M3 | 95.85 |
| | | M4 | 70.84 |
| | | M5 | 98.96 |
| 4 | Nimbawade, Dist.-Sangli | M2 | 97.28 |
| | | M3 | 99.81 |
| | | M4 | 93.23 |
| | | M5 | 98.36 |

### 5.6.5 Conclusions

The objective was to not only experimentally verify the cavitation phenomenon for microbial disinfection obtained in the simulations, but also to put together a Standard Operating Procedure (SOP) to guide pump users (common people) on how to obtain maximum disinfection.

From this case study, it was seen that computational fluid dynamics simulations could accurately predict the occurrence and the intensity of cavitation occurring in the check valve of the India Mark II hand pump and this was also successfully validated by the experimental results. It was proven that cavitation has micro-biocidal properties and the spherical shockwave and high velocity micro-jet have the capacity to render microbes unviable. Additionally, optimum conditions for disinfection were found to be 60% flow area opening as a function of the check valve. However, these dimensions may vary depending on the length of the lift and the water pumping rate (number of strokes).

Pilot plant experimental results corroborated the simulation results. Since cavitation is a stochastic phenomenon and it was proved to occur within the India Mark II hand pump, further low-pressure zones can be introduced to increase the probability of cavitation occurring. Thus, theoretically, 100% disinfection can be achieved. Moreover, the check valve modification can be applied to hand pumps in villages to disinfect water by 85–90%. Since it is reported that the structural integrity of the cell wall of pathogenic microbial matter is lower than *E. coli*, it can be stated that a large percentage of pathogenic matter will be killed.

## References

1. N. Nicomel, K. Leus, K. Folens, P. Van Der Voort and G. Du Laing, *Int. J. Environ. Res. Publ. Health*, 2015, **13**(1), 62.
2. W. Grabow, J. L. Slabbert, W. S. G. Morgan and S. A. A. Jahn, *Water SA*, 1985, **11**(1), 9–14.
3. A. A. Al-Anizi, M. Theresa Hellyer and D. Zhang, *Water Res.*, 2014, **56C**, 77–87.
4. J. P. Sutherland, G. K. Folkard and W. D. Grant, *Waterlines*, 1990, **8**(4), 30–32.
5. J. P. Sutherland, G. K. Folkard, M. A. Mtawali and W. D. Grant, Moringa oleifera as a natural coagulant, in 20th WEDC Conference, 1994.
6. D. C. Nguyen and M. Y. Han, *Resour., Conserv. Recycl.*, 2017, **126**, 219–227.
7. S. Saran, P. Arunkumar and S. P. Devipriya, *Water Sci. Technol.: Water Supply*, 2018, **18**(1), 49–59.
8. C. C. Amos, A. Rahman and J. M. Gathenya, *J. Cleaner Prod.*, 2018, **172**, 196–207.
9. K. Dwivedi and S. S. Bhadauria, *ARPN J. Eng. Appl. Sci.*, 2009, **4**(6), 31–38.
10. T. Missimer, R. Maliva, N. Ghaffour, T. Leiknes and G. Amy, *MDPI AG*, 2014, **6**(8), 2322–2338.
11. http://www.iwa-network.org/wp-content/uploads/2018/02/OFID-Wastewater-report-2018.pdf (last accessed july 2018).
12. S. Godfrey, P. Labhasetwar and S. Wate, *Resour., Conserv. Recycl.*, 2009, **53**(5), 287–293.
13. D. Ociński, I. Jacukowicz-Sobala, P. Mazur, J. Raczyk and E. Kociolek-Balawejder, *Chem. Eng. J.*, 2016, **294**, 210–221.
14. J. Mertens, J. Rose, B. Wehrli and G. Furrer, *Water Res.*, 2016, **88**, 844–851.
15. S. Ahoranta, M. Kokko, S. Papirio, B. Ozkaya and J. Puhakka, *J. Hazard. Mater.*, 2016, **306**, 124–132.
16. M. U. Sankar, S. Aigal, S. M. Maliyekkal, A. Chaudhary, A. A. Kumar, K. Chaudhari and T. Pradeep, Biopolymer-reinforced synthetic granular nanocomposites for affordable point-of-use water purification, *Proc. Natl. Acad. Sci. U. S. A.*, 2013, **110**(21), 8459–8464.
17. V. Vaiano, G. Iervolino, D. Sannino, L. Rizzo, G. Sarno and A. Farina, *Appl. Catal., B*, 2014, **160**, 247–253.
18. L. Handia, J. M. Tembo and C. Mwiindwa, *Phys. Chem. Earth*, 2003, **28**, 893–896.
19. E. A. Tobin, T. F. Ediagbonya, G. Ehidiamen and D. A. Asogun, *J. Public Health Epidemiol.*, 2013, **5**, 479–487.
20. A. Guebaili, Y. Djebbar, A. Guedri and S. Boukhari, *J. Mater. Environ. Sci.*, 2011, **2**(S1), 427–432.
21. I. Adler, K. A. Hudson-Edwards and L. C. Campos, *J. Water Supply: Res. Technol.*, 2013, **62**(8), 545–551.
22. M. Lee, M. Kim, Y. Kim and M. Han, *J. Environ. Manage.*, 2017, **200**, 400–406.

23. https://www.unicef.org/about/annualreport/files/Communication_AR_2010.pdf (UNICEF 2010).
24. https://www.unicef.org/supply/files/UNICEF_Supply_Annual_Report_2014_web.pdf (UNICEF 2014).
25. V. K. Saharan, D. V. Pinjari, P. R. Gogate and A. B. Pandit, *Industrial Wastewater Treatment, Recycling and Reuse*, 2014, pp. 141–191.
26. T. Sivasankar and V. S. Moholkar, *Chem. Eng. J.*, 2009, **1–3**(149), 57–69.
27. M. P. Badve, T. Alpar, A. B. Pandit, P. R. Gogate and LeventeCsoka, *Ultrason. Sonochem.*, 2015, **22**, 272–277.
28. V. K. Saharan, A. B. Pandit, P. S. Satish Kumar and S. Anandan, *Ind. Eng. Chem. Res.*, 2012, **51**, 1981–1989.
29. K. Jyoti and A. B. Pandit, *Biochem. Eng. J.*, 2001, **7**, 201–212.
30. http://www.hycator.com/tech.html#.
31. M. P. Badve, M. N. Bhagat and A. B. Pandit, *Sep. Purif. Technol.*, 2013, **106**, 15–21.
32. T. Sivasankar and V. S. Moholkar, *Ultrason. Sonochem.*, 2009, **16**, 769–781.
33. S. T. L. Harrison and A. B. Pandit, *The Disruption of Microbial Cells by Hydrodynamic Cavitation*, Washington, 1992.
34. S. S. Save, A. B. Pandit and J. B. Joshi, *Chem. Eng. J.*, 1994, **55**, B67–B72.
35. S. S. Save, A. B. Pandit and J. B. Joshi, *Trans. Inst. Chem. Eng.*, 1997, **71C**(75), 41–48.
36. B. Balasundaram and A. B. Pandit, *J. Biochem. Eng.*, 2001, **8**, 251–256.
37. M. Sivakumar and A. B. Pandit, *Ultrason. Sonochem. J.*, 2002, **9**(3), 123–131.
38. S. Manchalwar, Disinfection of potable water using hydrodynamic cavitation, Ph.D. thesis (Institute of Chemical Technology), 2015.

CHAPTER 6

# Selection Criteria of Water Treatment Strategy in Developing Countries

Having read the book so far and understanding the different water treatment methods available for developing nations, it would naturally be interesting to note the major techno-economic and social differences between the various methods discussed in this book. Having understood the options, it is necessary to narrow down the choices of a suitable water treatment process for a developing country scenario. Therefore, this chapter mainly aims at bringing out the salient features of each technique of water treatment based on some major criteria like the mechanism of action, pollutants removed, major advantages and disadvantages, cost and the scale of operation.

## 6.1 Introduction

When one must select an appropriate methodology to treat drinking water under the constraints imposed in a developing country, it must be based on certain selection criteria. Such selection standards can be many, such as the source water being treated, the types of contaminants, the efficiency and economics and the affordability of the selected process *etc.* Some factors such as the cost and scale of operation become very important, especially when water treatment is aimed at developing nations where usually affordability is a major factor.

This book has described many methods that are particularly applicable for treating water in developing countries. All the techniques referred to in this

Drinking Water Treatment for Developing Countries: Physical, Chemical and Biological Pollutants
By Aniruddha Bhalchandra Pandit and Jyoti Kishen Kumar

Published by the Royal Society of Chemistry, www.rsc.org

book are compared based on treatment mechanism of action, extent of contaminants eliminated, important merits and demerits, economics and the scale of operation (Table 6.1). As discussed in the earlier chapters, it is important to once again emphasize here that there is no one universal method that will solve the water contamination challenge in developing nations.

Based on the specific needs of a developing country and considering the quality of source water, impurities involved, cost of operation and the overall pros and cons, a water treatment method may be suggested. Here, the volume of water to be treated also needs careful consideration. Some suggestions for possible and potential drinking water treatment techniques based on pollutants, scale of treatment and cost are discussed below and shown in Figure 6.1. These are only indicative and reflect the authors views, they may not necessarily always apply and variations are possible on a case to case basis.

## 6.2 Selection of Water Treatment Technique Based on Pollutants

Water may contain different physical, chemical and biological pollutants based on its source, geographical location and the industrial or human activities in the area and this has already been discussed at length in the previous chapters. This section highlights some water treatment techniques that may be specifically used to remove certain specifics types of pollutants.

Under each category of pollutants (physical, chemical and biological), there are a plethora of impurities that may arise in the source water. It is beyond the scope of this book to state a water treatment method for each contaminant under the three classes. However, an attempt has been made to suggest possible methodologies for the specific class of contaminants.

Based on the previous chapters and Table 6.1, it can be concluded that most physical, chemical and biological pollutants can be removed by most water treatment techniques described in this book (which specifically targets developing countries). For instance, chlorination and some conventional filtration systems such as activated carbon filters, and natural and hybrid filters are well suited to remove any or all of these pollutants from any contaminated water source. Similarly, among the emerging techniques, the solar ball, water purifying bicycle, life sack, AWGs, hydrodynamic cavitation and pure water bottles are equally efficient at eliminating most of the pollutants on an individual or a smaller scale of operation.

It is interesting to note that among all the filtration methods, only slow sand filtration can be used for the treatment of most physical, chemical and biological pollutants including viruses and protozoan cysts. Rapid sand filtration and ceramic filters are also useful for most contaminants, but they are not particularly useful to eliminate viruses and are able to remove only some viruses from the polluted water.

**Table 6.1** Comparison of different water treatment methods suitable for developing countries.

| S. no. | Method | Mechanism of action | Pollutant treated | Merits | Demerits | Treatment cost | Scale of operation (litres per day)[a] |
|---|---|---|---|---|---|---|---|
| 1 | Conventional SODIS | UV rays of the sun resulting in reactive oxygen species that kills microorganisms | Bacteria, Fungi and Protozoa | Low cost, no electricity, ease of use | Ineffective in turbid waters, viruses are resistant | US$0.5 $m^{-3}$ [b] [2] | Household to community level Bottle size of 0.5 to 2 L and 20–50 L |
| 2 | Thermal enhancement of SODIS | Optical & thermal effects of solar rays | Bacteria, Fungi and Protozoa | Reduction of treatment time | Not effective on cloudy days | Additional cost of US$0.04 $m^{-3}$ for treatment using natural additives | Higher cost of additives & enhancements can be justified only if the volume of water treated is large or if the water quality after treatment is high |
| 3 | SODIS & photo-catalysis | $TiO_2$ used increases the formation of ROS species | Bacteria, Fungi and Protozoa, viruses, *Cryptosporidium parvum* oocyst, *Clostridium perfringens* spores | Reduction of the irradiation dose or treatment time | Post treatment recovery of suspended titanium dioxide as it may lead to taste and health issues if present in treated water | US$2.6–8 $m^{-3}$ for system with production capacity from 0.04 to 0.1 $m^3$ per day per $m^2$ of solar collector depending on location[3] <US$1 $m^{-3}$ for flow-through pasteurization system of capacity 200 L $m^2$ per day. | |
| 4 | SODIS and additives | Chemical or natural additives augment SODIS by enhancing ROS species formation | MS2 coliphages and several Enterococcus species, protozoa cysts | Reduction in treatment time to 1–2 h | Chemical additives are expensive and natural ones may lead to toxicity | | |
| 5 | Augmenting SODIS with bottle/reactor modification | The modification helps in increasing the intensity of solar rays | *C. parvum* oocysts, Total HPC bacteria | Reduction in treatment time to 3–4 h & higher inactivation rates | Availability of materials and added maintenance | | |
| 6 | Chlorination | Reacts with cell membrane & affects vital processes | Bacteria, most viruses | Most common method, easy to apply, low cost method | Many Protozoa are resistant. Formation of disinfection byproducts such as THM, which is carcinogenic | US$2–5 $m^{-3}$ for domestic system US$1–3.75 $m^{-3}$ for 100 MLD <US$0.05 $m^{-3}$ for >100 MLD systems | Domestic, community and large-scale treatment |

| | | | | | | | |
|---|---|---|---|---|---|---|---|
| 7 | Slow sand filtration | Removal by action of biological layer formed on the filter media | Turbidity, organic and inorganic contaminates, bacteria and some protozoan oocysts such as *Giardia* and *Cryptosporidium* | Sand is easily available, simple gravity-based process, & hence no electricity needed | Pre-treatment needed to prevent clogging & disinfection post filtration may be needed sometimes. Cannot remove all pathogenic bacteria & viruses | The Kanchan Filter based on sand filtration costs US$15–70 for a household unit. It has a treatment cost of 1.2 US$ $m^{-3}$.[2] The average construction cost ranges from US$15–US$60 & cost per litre of treated water is 0.068 US cents.[5] | Used for treating surface water in small rural communities where available land is not a limiting factor for slow sand filtration |
| 8 | Ceramic filters | Particles larger than the pores (usual pore size of filter used is about 0.2 μ) get accumulated on the filter surface | Total coliforms, *E. coli*, fecal coliforms and a host of protozoa and some viruses | Simple and easily available material used, produced by local population | Sustainability, community compliance and maintenance of the filter | Cost per liter treated (including cost of filter only) is 0.034–0.14 US cents.[5] US$2.2–5 $m^{-3}$ for household use. | Household level & pilot scale |
| 9 | Rapid sand filtration | Homogeneous oxidation, heterogeneous oxidation and biological oxidation | Iron, manganese, most bacteria, protozoa, a few viruses | Faster process and requires less land compared to slow sand filtration | Pre-treatment and post disinfection needed, frequent filter maintenance & high energy required. | Higher than slow sand filtration. Cost is double that of SSF.[6] | Used for Municipal treatment levels. May not be suited for small communities in developing countries owing to the higher cost as compared to slow sand filtration. |
| 10 | Activated carbon filtration | Physical absorption | Most physical, chemical & biological pollutants | Ease of availability, high surface area that is amenable to modifications | Coal, wood *etc.* Cannot be used as source of carbon due to global deforestation issues, due to collection of biological impurities on long term use, it may lead to clogging or lead to further contamination. | Most filters cost between US$8 to US$10. 10 cents to US$1.00 per 1000 gallons of water.[7] | Can be used from household scale to large scale water purification. |

**Table 6.1** (*Continued*)

| S. no. | Method | Mechanism of action | Pollutant treated | Merits | Demerits | Treatment cost | Scale of operation (litres per day)[a] |
|---|---|---|---|---|---|---|---|
| 11 | Natural filters (wood, plant xylem) | Physical absorption due to high porosity and large surface area | Most bacteria, physical and chemical pollutants | Locally available, economical, no power consumption, biodegradable | Filtration efficiency depends on porosity, density of wood | Around US$15 as initial investment for making xylem filters | Mostly household level, in some cases pilot scale has been investigated. |
| 12 | Hybrid filters | Combination of two or more mechanisms (*e.g.* Filtration+ germicidal action of metals like silver) | Most biological, physical and chemical pollutants | Improved efficiency due to enhanced removal of pollutants by synergistic effects of the combination used | Expensive compared to conventional filtration, clogging of filters may require periodic maintenance | Terafil – around US$0.5 for the filter and approximately US$4 for the entire unit. Tata Swach® – running cost of US$1.5 $m^{-3}$ and a capital cost of US$50 | Mostly household or community level |
| 13 | Herbal approaches (*e.g.* Tulsi, *Moringa*, neem *etc.*) | Active ingredients have coagulating and anti-microbial properties | Most biological, physical and chemical pollutants. Removes turbidity. | Readily available as many medicinal plants are locally found in developing countries, inexpensive, easy to apply | Toxicity and safety of the herbs must be ascertained for long term effects. | Cost is negligible as the herbs are locally cultivated. | Household level. Can be used for large scale purposes also. |
| 14 | Desalination using conventional energy sources | Heat in case of thermal processes and semipermeable membranes form barriers for salt in case of membrane processes. | Removes salt from sea water and results in pure water fit for drinking. | Vast water bodies in the form of seas are available across the globe, ancient and established method | Requires huge amount of energy, expensive, emission of greenhouse gases due to use of conventional energy sources | Very expensive due to high installation costs and huge energy demand (*e.g.* US$89 000 for RO based plant in Tuvalu, Pacific Island, 2006) | Not suitable for developing countries where the economic conditions are poor; GNI per capita of $1025 or less.[8] |
| 15 | Desalination using renewable energy sources | Solar, wind, geothermal or wave energy used to power the plants. | Removes salt from sea water and results in pure water fit for drinking | Relatively inexpensive due to the use of renewable energy sources that power the plants | Can be adopted preferably only by those developing countries having solar energy, geothermal spots, | Cheaper than desalination by conventional energy sources (*e.g.* cost of geothermal plant in | Appropriate for developing countries on community and large scale |

| | | | | | | | |
|---|---|---|---|---|---|---|---|
| | | | | | wave power (being close to sea) and/or high wind potential | Queensland is US$0.73–1.46 $m^{-3}$) | |
| 16 | Drinkable book | Made of around 20 pages of specialized filter paper (Silver nanoparticles coated on cellulose fibers of an absorbent blotting paper sheet) | *E. coli* and *E. faecalis* | Easy to use, simple technology, educational element (water safety tips & precautions are printed on filter paper in English and the local language) | Cost of silver, scale up issues, possible clogging of the filter paper, reusability and leaching of silver during long term | Each page costs only about 10 cents and it can treat about 100 L of water.[9] | Household level |
| 17 | Lifestraw | Hollow fibers (pores <0.2 microns) present inside the filter body trap the pathogens. | 99.99% of bacteria such as *E. coli, Vibrio cholerae* and *Salmonella typhi* and 99.9% of protozoan cysts like *Cryptosporidium parvum* and *Giardia lamblia* | No electricity needed, simple to use and handle | Clogging of filters, replacement of device | Around US$15 for treating 1000 L of water.[10] | Individual & family use |
| 18 | Solar ball | Solar evaporation | Most biological, physical and chemical pollutants | Spherical shape, compact & unique design, captures solar rays from 360 degrees, easy to handle | May be suitable only for regions that receive high levels of sunlight, scalability may not be easy | Approximately US$35 to US$50 per unit and can produce up to 3 L of clean water per day.[11] | Stage of prototype |
| 19 | Water purifying bicycle | The water purification system (consisting of filters) attached on the bicycle is operated by simply pedaling the bicycle | Most biological, physical and chemical pollutants | Mobile and can be taken to inaccessible areas, no engine or battery is needed, suitable for remote places & in emergency situations like natural disasters | Filters used may become clogged, relatively expensive for developing countries | It is currently priced at about US$6600 & can filter 5 L of water per min & efforts underway for price reduction.[12] | Community level |

**Table 6.1** (*Continued*)

| S. no. | Method | Mechanism of action | Pollutant treated | Merits | Demerits | Treatment cost | Scale of operation (litres per day)[a] |
|---|---|---|---|---|---|---|---|
| 20 | Life sack | Combined effect of solar rays that fall on a plastic bag made of PVC + an embedded filter (15 nm) results in pollutant removal. | Most microorganisms | Simple, easy to use and carry, the life sack serves the dual role of providing food supplies & pure water to remote areas. | The sack can be used only after the food supplies are over, compliance from the end user is very essential, may not work for regions with less sunlight, filter system can get clogged. | Low priced (exact value not available) | Suitable for a family (around 20 L) |
| 21 | Atmospheric Water Generators (AWGs) | The device extracts the humidity present in the air and delivers purified water (after filtration & UV treatment). | Most biological, physical and chemical pollutants | Good option when other sources of water (lakes, municipal supply, wells) are not available, eliminates need to travel for water | Huge energy costs, the device is also expensive | Cost varies depending on the company, suppliers *etc.* For Ecoloblue Product, the price is around US$800 to US$1500 depending on the model to produce up to 30 L per day.[13] | Can be used on a large scale. However, may be expensive for use in developing countries |
| 22 | Pure water bottle | Dual chamber compact bottles with filters and UV | Most biological, physical and chemical pollutants | Simple, compact, easy to carry during travels, takes less time to get pure water (about 2 mins) | Filters can result in clogging and UV system may need maintenance, cannot hold large amount of water. | Exact cost not available | Still in prototype stage |

[a]Household Level = approximately 20 L per day; Community Level = approximately 20 000 L per day; Large Scale Level means >100 MLD
[b]Assuming 100 uses of PET bottles before they cannot be used further.

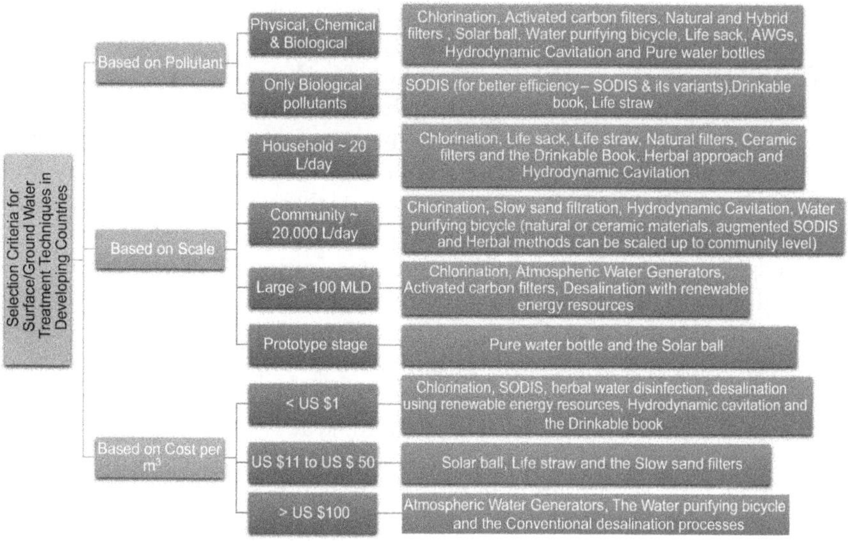


**Figure 6.1** Selection criteria for surface/ground water treatment techniques in developing countries.

Methods such as SODIS and its variants are useful only when biological pollutants must be inactivated. They cannot eliminate or remove physical and chemical contaminants. Even under the class of biological contaminants, it must be emphasized that only bacteria can be inactivated by the conventional SODIS method. For the removal of viruses and protozoal cysts, SODIS and photocatalysis or additives and some form of augmentation (such as reactor modifications) of SODIS are required. For treating biological contaminants such as bacteria, even emerging methods such as the drinkable book can be considered. Lifestraw is also a good option, especially when viruses and protozoal cysts need to be eliminated. In general, removal of physical impurities is relatively easy and cheap, followed by chemical impurities, and the removal of biological impurities is most difficult. Also, the physical and chemical impurities may show a long-term effect if not treated whereas the effect of biological contaminants may be immediate.

## 6.3 Selection of Water Treatment Technique Based on Scale of Operation

The human need for safe and pure drinking water is perennial and will continue to be so. However, the requirement may vary, *i.e.* safe and clean drinking water demand can differ from individual needs to huge volumes for larger numbers of people. This is true for any region including developing countries and therefore drinking water treatment schemes should be able to meet these demands and provide potable water from a small to large scale.

Numerous water treatment methods described here and compared in Table 6.1 are available for addressing the varied requirements of potable water in developing countries. For individual and/or household scale, treatment methods based on filters such as the Life sack, lifestraw, natural filters, ceramic filters and the Drinkable Book are suitable. The herbal approach could also be considered as an alternative. For community level, slow sand filtration and, among the newer techniques, the water purifying bicycle and hydrodynamic cavitation appear to be promising.

Atmospheric water generators and conventional methods such as activated carbon filters are excellent for treating larger volumes of water. The desalination process can also be considered for large scale water treatment, especially if it is powered by renewable energy sources such as solar, wind, wave or geothermal methods, as described in this book.

Among the methods described in this book, it is interesting to note that there are some techniques that can be scaled up and therefore would cater to both small and community level requirements of safe potable water. Examples of such methods are chlorination and herbal water disinfection that can be used for household level and can be scaled up to pilot and large scale. The *Moringa oleifera* case study in chapter five is an apt example for this situation. Yet another methodology that falls in this category is filtration, especially that based on natural or ceramic materials, and hydrodynamic cavitation, as they can be scaled up to community levels. Moreover, augmented SODIS techniques where the effect of SODIS is enhanced due to photocatalysis, additives and reactor modifications as described earlier are also very apt for scaling up to community levels.

Fascinatingly, among the methods reviewed in this book, it appears that other than chlorination, the only filtration method that can be effectively used from household level to community and larger scale is the activated carbon filtration technique. Here, it is important to realize that some emerging and novel methods such as the pure water bottle and the solar ball are still in the prototype stage and with increased research and support from various stakeholders, it will not be surprising to find these techniques evolving into larger scale versions in the times to come.

Community level treatment systems are found to be suitable when local drinking water resources are contaminated with specific pollutants such as arsenic and fluoride, provided they are maintained regularly.

## 6.4 Selection of Water Treatment Technique Based on Cost

For any water treatment to be effectively deployed under the constraints in a developing country, it must be low-cost, and the importance of considering the affordability aspects has been highlighted in several sections of this book. Thus, cost becomes one of the major criteria for selecting a water treatment technique for low income countries where the people do not have access to basic utilities such as electricity and there is a lack of funds for the purchase of expensive devices and chemicals routinely used for water treatment.

When cost considerations are applied to the treatment methodologies discussed in this book, it is very evident that methods such as chlorination, SODIS, herbal water disinfection, desalination using renewable energy resources and the drinkable book (among the emerging methods) are appropriate for deploying in developing countries. This is mainly because the treatment cost of these methods is either negligible or less than US$1 $m^{-3}$. In a developing country, when community water supply (Municipality) itself is about 5 to 25 cents per $m^3$, individual treatments should be marginally more expensive. Tata swatch costs about INR 300 (cartridge cost) for 3000 (3 $m^3$) l of water, meeting WHO standards, *i.e.* 1.5 $m^{-3}$. The cost of hydrodynamic cavitation for treatment of lake water described in case study 5 was around 0.25 cents per $m^3$, which is again a viable option.

At the next slightly more expensive level, activated carbon filters and ceramic filters appear promising as their cost can vary between less than US$1 $m^{-3}$ to about US$10 $m^{-3}$. Many hybrid and natural filters also fall in this range, but their cost may go up to US$50 $m^{-3}$ and more depending on the combinations used to create the hybrid filter and the cost of natural fibers in the case of the natural filters.

Other water treatment methods that cost between US$11 and US$50 are the solar ball, lifestraw and slow sand filters. It is important to note that methods such as atmospheric water generators, the water purifying bicycle and the conventional desalination processes are expensive and cost more than US$100 $m^{-3}$. Therefore, these methods can be chosen only if there is sufficient funding available or if these solutions can be afforded by the consumer as the initial capital expenditure itself is high.

## 6.5 Selection of Water Treatment Technique for Harvested Rain Water

From the previous chapters, it can be clearly understood that apart from treating existing water sources such as ground and surface water, it is also very critical to conserve water resources by ways such as rain water harvesting and waste water reuse. The preceding sections of this chapter discussed the selection criteria for the best treatment techniques for surface and/or ground water that can be used for piped or household drinking water purposes. Extending the same criteria for the selection of possible treatment methods for treating harvested rain water has been presented in this section.

It is important to recall that several treatment methods for treating harvested rain water and the cost comparisons have been elaborated in Chapter 3. However, these economic comparisons were for the overall RWH set-up that included the storage tanks, piping and other infrastructure materials needed. The main objective in this section is to compare only the cost of the treatment methods used routinely for treating harvested rain water. Table 6.2 lists and compares various methods based on their

**Table 6.2** Comparison of different water treatment methods suitable for treating harvested rain water in developing countries.

| S. no | Method | Mechanism of action | Pollutant treated | Merits | Demerits | Treatment cost | Scale of operation[a] |
|---|---|---|---|---|---|---|---|
| 1 | Chlorination | Reacts with cell membrane & affects vital processes | Bacteria, most viruses | Most common method, easy to apply, low cost method | Many protozoa are resistant. To be applied after removal of rain water from storage tank to prevent reaction with organic matter | US$2–5 $m^{-3}$ for domestic system US$1–3.75 $m^{-3}$ for 100 $m^3$.[4] < US$0.05 $m^{-3}$ for > 100 MLD systems | Domestic, community and large-scale treatment |
| 2 | Slow sand filtration | Removal by action of biological layer formed on the filter media | Can reduce most bacteria & some protozoa | Economical, sand is easily available, simple gravity-based process | Cannot remove viruses. Constant water flow needed to be effective | Average cost: US$15–$60 & cost per litre of treated water is 0.068 US cents.[5] | Community level |
| 3 | SODIS | UV rays of the sun resulting in reactive oxygen species that kills microorganisms | Bacteria, Fungi and Protozoa | Low cost, free natural resource available in developing countries. Can be enhanced by reflectors | Cannot be used if concentration of suspended solids is >10 $mg\,L^{-1}$. Viruses are resistant | US$0.63 per person per year | Household to community levels |
| 4 | Membrane filtration | Physical removal based on pore size of (0.1 μm) | Bacteria, Fungi and Protozoa | Efficient process | Needs maintenance, expensive, cannot remove viruses | Higher cost based on type of membrane | Household to community levels |
| 5 | Rotating disc filter with ceramic membrane | Physical removal based on pore size of (0.06 μm) | Bacteria and Viruses | Smaller pore size, better removal efficiency | Needs maintenance, Expensive | Expensive | Suitable for central treatment of rain water |
| 6 | Filtration + Adsorption + Disinfection | Removes turbidity and organic matter and disinfection kills microbes | Turbidity, organic matter, microorganisms | Hybrid process and therefore more effective | No disinfectant residue therefore regrowth of microbes is possible | Higher cost due to hybrid process | Household to community levels |
| 7 | Ultraviolet radiation | Reacts with molecular bonds within DNA of microorganisms | Bacteria and protozoa | Easy to install, operate and maintain | Water must have low turbidity, lamps need periodic cleaning/replacement | 7 cents per cubic meter for 100 $m^3$ per day.[4] | Household to community levels |

[a]Household Level = approximately 20 L per day; Community Level = approximately 20 000 L per day; Large Scale Level means >100 MLD.

mechanism of action, contaminants eliminated, important merits and demerits, economics and scale of operation.

For developing countries where low cost and efficient simple methodologies are the most feasible ways to get safe and clean harvested rain water, the first treatment of harvested rain water can be done by using the first flush water diverters. This method has been elaborated in Chapter 3 and is an efficient way to prevent the entry of several contaminants especially coarse particles from the harvested rain water. Moreover, it acts like a pre-treatment and the water that has passed through a first flush device would have a lower pollutant load and hence requires less treatment. The overall cost of the treatment also gets reduced substantially. After this stage, based on either the type of pollutants, scale of operation or the cost, the following strategies may be employed for selecting an appropriate method of treatment.

Based on the pollutants to be treated, it is clear from Table 6.2 that the hybrid methods are good if physical, chemical and biological contaminants are to be removed. This is mainly because, in a hybrid method, the combination of treatment technologies employed, such as filtration, adsorption and finally disinfection, can eliminate a large variety of pollutants from harvested rain water. When only microorganisms must be eliminated, SODIS is a good choice, and for specifically viral inactivation, the rotating disc filter with a ceramic membrane appears to be the best choice due to the small pore size of 0.06 μm or less.

If the scale of operation must be considered, then chlorination is the first and best choice for any level of operation from household to community and large-scale levels. For household to community levels, methods such as SODIS, membrane and hybrid filters and UV treatment can be employed. For a central treatment of rain water, the rotating disc filter with a ceramic membrane is a good option and slow sand filtration can be considered for only community level operation where the availability of land may not be an issue.

Cost efficient techniques appear to be UV treatment, SODIS and chlorination as these treatments may cost anywhere from 7 cents to US\$3.75 $m^{-3}$. The membrane and hybrid processes are expensive, and the cost may vary based on the type, pore size and material of the filter media or membrane. Additionally, the hybrid techniques involve more than one method, thus leading to augmentation of the initial capital cost. Therefore, these methods may be useful only when the highest quality of treated rain water is required, thereby justifying the cost of treatment. Slow sand filtration can be considered as moderate in cost (US\$15–US\$60 $m^{-3}$) and can be applied to those developing nations that can afford the cost and can operate on a larger community scale (Figure 6.2).

## 6.6 Selection of Water Treatment Technique for Waste Water Reuse

Waste water reuse is yet another water conservation technique that can be a promising source of water provided it is treated and made fit either for

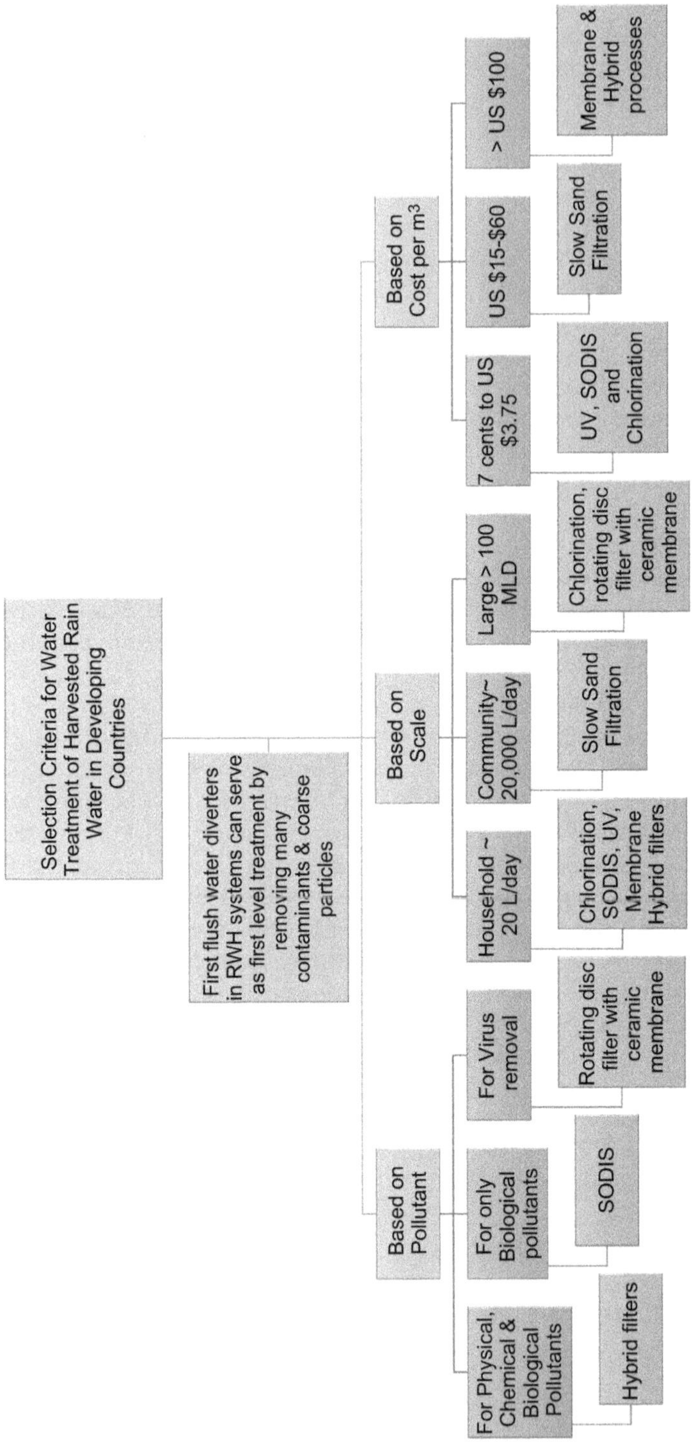


**Figure 6.2** Selection Criteria for Water Treatment of Harvested Rain Water in Developing Countries.

non-potable or potable use depending on the end user. A lot has been elaborated in Chapter 4, where indirect and direct potable reuse has been discussed in detail in the context of developing countries. Based on these details, it can be stated that these technologies are most suitable for large scale use (25 to 120 MLD) where the funds needed for such methodologies can be raised through a common pool and deployed effectively (Figure 6.3).

Thus, these are relatively expensive methods with operating costs of about US$53 million per year for DPR and US$72 million per year for IPR systems for a treatment capacity of 120 million litres per day (MLD).[1] Between the two, IPR systems are more expensive and rely on enclosed pipelines and pumping systems over long distances. On the other hand, DPR appears to be lower in cost compared to IPR systems, as drinking water treatment and waste water treatments typically occur at nearby locations, therefore DPR systems require less pumping of water to reach the end consumers.

Based on the pollutants, these treatment systems are effective as they employ a treatment train involving membrane-based systems that employ microfiltration or ultrafiltration, reverse osmosis, UV or an advanced oxidation process. Thus, most physical, chemical and biological contaminants are safely removed. However, safety issues are present with both these systems, especially DPR because it is intended for direct consumption by the

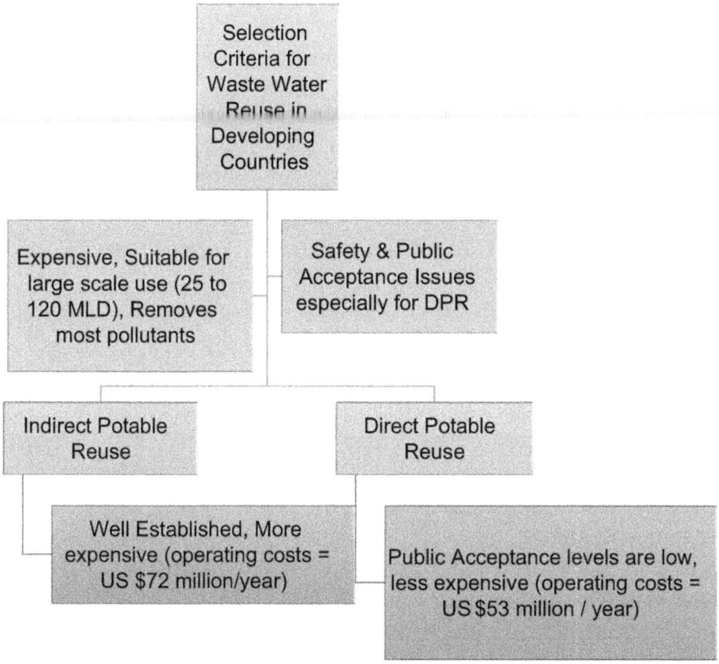


**Figure 6.3** Selection Criteria for Waste Water Reuse in Developing Countries.

user. Thus, the ability of DPR systems to fully remove certain compounds such as pharmaceutical residues, residues of personal care products and certain endocrine disrupting compounds from water is a major factor to be considered when selecting this option.

Overall, IPR systems are relatively well established globally, and presently, their public acceptance is also higher as compared to DPR. But, when it comes to direct drinking purposes, public acceptance is still a major hurdle in the implementation of DPR projects. However, with increasing public education and outreach programs, DPR could become a commonly adopted water conservation method in the coming decades.

## 6.7 Conclusions

The best possible water treatment technology suitable for developing countries can be selected based on several criteria such as the source water to be treated, pollutants to be removed, scale of operation and the cost. With the various treatment methods discussed in this book for developing nations, it can be concluded that the best choice of treatment method may vary depending on whether the raw water is from a surface or ground source or is conserved water either from harvested rain water or reused waste water. For both cases, the options are different and are based on several factors like the type of contaminants involved, and whether it is needed on household, community or a large scale. Finally, the economics of the treatment method is an important criterion as low-cost methods are most suited for developing nations. Thus, appropriate water treatment methods as suggested in this chapter can be used as potential means for providing safe drinking water for the population residing in developing countries.

## References

1. https://www.atse.org.au/ (last accessed July 2018).
2. A. K. Plappally and J. H. Lienhard V, *Desalin. Water Treat.*, 2013, **51**(1–3), 200–232.
3. N. Dainelli, G. Manfrida, K. Petela and F. Rossi, *Energies*, 2017, **10**(9), 1395.
4. M. H. Dore, R. G. Singh, A. Khaleghi-Moghadam and G. Achari, *Int. J. Water Resour. Environ. Eng.*, 2013, **5**(2), 100–109.
5. https://www.cdc.gov/safewater/sand-filtration.html (last accessed July 2018).
6. https://www.ircwash.org/sites/default/files/255.1-87SL-3062.pdf (last accessed July 2018).
7. https://www.wateronline.com/doc/the-real-cost-of-treating-drinking-water-with-0001(last accessed July 2018).
8. https://data.worldbank.org/products/wdi-maps (last accessed July 2018).
9. https://food.ndtv.com/food-drinks/low-cost-paper-based-method-to-purify-drinking-water-1208046 (last accessed July 2018).

10. https://www.lifestraw.com/products/lifestraw/(last accessed July 2018) (last accessed July 2018).
11. http://earthtechling.com/2011/03/solar-powered-ball-purifies-dirty-water/ (last accessed July 2018).
12. https://www.wired.com/2011/02/bike-purifies-water-with-pedal-power/ (last accessed July 2018).
13. https://ecoloblue.com/atmospheric-water-generator/228-148-ecoloblue-30e.html (last accessed July 2018).

CHAPTER 7

# Conclusions

## 7.1 Summary

A topic such as drinking water and its treatment that is so important to human life and its survival cannot be concluded in a few pages or documented in just one book. This is mainly because it is a vast field and is particularly relevant in current times due to the ever-increasing water demand globally and severe stress on water resources worldwide. Needless to state, this is magnified in all developing countries, where individuals have a low earning power. A humble attempt has been made to put together most of the available information dealing with drinking water treatment and its conservation that is especially relevant to developing nations. Based on all the chapters and topics covered and the methods and strategies discussed, the following summary conclusions can be drawn. Each of the topics discussed briefly here has the capability of getting converted into a separate book, however, in this book, an attempt has been made to at least quantitatively address most of the major and minor methods.

Safe and clean drinking water is of paramount importance to prevent health risks and this is even more relevant to developing countries as most of these regions do not have access to basic clean water resources and/or have polluted water bodies due to sewage and many other forms of contamination. From the WHO report (described in chapter 1), it is evident that a huge population of almost 663 million lack access to basic quality drinking water and many who have access invariably end up consuming contaminated water. Therefore, one of the main agendas is to increase international collaboration to support developing countries in water treatment and management related activities that include water treatment, reuse and water harvesting. Moreover, there is a huge difference in services between urban and rural regions and hence there is an urgent need to improve access to

Drinking Water Treatment for Developing Countries: Physical, Chemical and Biological Pollutants
By Aniruddha Bhalchandra Pandit and Jyoti Kishen Kumar

Published by the Royal Society of Chemistry, www.rsc.org

safe and clean drinking water in the rural regions, which are mainly situated in developing countries across the world.

Physical, chemical and biological pollutants are of countless varieties and can get introduced into source water by various means. Emerging pollutants such as pharmaceutical compounds, fertilizers, pesticides and their derivatives have started becoming a major threat to human health. Many of these substances are known to act as endocrine disruptors (EDCs), which change the normal functions of hormones resulting in several heath risks and severe acute toxicity issues. Therefore, treatment suited to each class of pollutant is an important yet challenging task. Several water treatment techniques have been adapted that remove most of the physical, chemical and biological contaminants as discussed in this book. However, novel techniques such as those described here and similar methods yet to be invented, and in the pipeline, offer a ray of hope to get safe and clean potable water.

There is no one universal water treatment technique for developing countries and the treatment method should be selected based on various criteria such as the origin of the source water, pollutants to be removed, scale and efficacy of operation and the cost (described in chapter 6). Waste Water Reuse technologies are most suitable for large scale use (25 to 120 MLD) and are very expensive methods if rain water harvesting is not possible. Therefore, it can be used in those regions where the funds needed for such methodologies can be raised and deployed effectively along with extreme scarcity. Moreover, the IPR systems are relatively well established globally and accepted as compared to DPR. Public acceptance is still a major obstacle in the implementation of DPR but due to the severe water crunch and increase in public awareness, DPR could emerge as one of the adopted and acceptable water conservation methods in the times to come. Rain water harvesting will continue to remain one of the most commonly used methods to conserve water in developing countries and simple first flush devices and hybrid methods can be easily used to treat harvested rain water to make it potable.

## 7.2 Recommendations for Future Work

Individual chapters in this book have recommendations listed at the end to enable the reader to understand the possible future work under each of the topics discussed. Overall recommendations are listed here at the end of the book for better clarity on the path forward. They are as follows:

- There is a need for more research and effort on the development of low-cost effective water treatment technologies suited for developing countries for the times to come.
- Development of customized treatment technologies that can specifically eliminate a class of physical, chemical or biological pollutants is also needed rather than a general overall scheme.
- Efforts to ensure that these technologies reach the end user in developing countries are the need of the hour. Moreover, ensuring

compliance by the consumer and dissemination of required education, awareness and knowledge are also required.

- Emphasis on the conservation of water and improving the quality of conserved water by rain water harvesting and waste water reuse in developing countries will be vital in the decades to come.

## 7.3 Ongoing Initiatives

To conclude this book, some positive initiatives occurring globally to meet the drinking water challenge are presented in this section. One among them is Nature based solutions (Biomimicry) for Water treatment, a report that was launched by the United Nations in March 2018. Nature based solutions involve the use of natural processes to augment water availability, improve water quality and reduce risks related to water borne diseases and disasters. It is important to note that currently, water management is carried out by conventional human built infrastructure and therefore there is a huge potential to integrate nature-based solutions such as green infrastructure that is cost effective and can work in tandem with the existing systems. For instance, eco-friendly forms of water storage, such as natural wetlands, improvements in soil moisture and more efficient recharge of groundwater, could be more sustainable and cost-effective than traditional grey infrastructure such as large dams. Yet another major challenge is the quality of water and it is well understood that proper source water protection can substantially reduce water treatment costs. Soil, crops, forests and grasslands can play an important role in maintaining water quality by trapping of the pollutants during storage or distribution, recycling of the nutrients and reducing sediment loading in water. Finally, nature-based solutions during natural disaster such as floods can involve water retention by managing infiltration and overland flow and floodplain restoration (http://www.unwater.org/publications/world-water-development-report-2018/).[1]

In a very useful initiative by UNICEF, the people in the villages of Kunia and Chamba in Malawi were provided solar powered water pumps for pumping water into storage tanks. Solar power enabled deeper drilling underground that enables reaching ground water table levels even during a drought when the water table drops. These solar powered water pumps have low maintenance and can last for more than a decade. Moreover, they are cheaper, sustainable and environmentally friendly compared to diesel generators. This is an apt example of a simple yet significant solution to water challenges faced by many developing countries that can lead to other improvements such as better school attendance, as children and women do not have to spend long hours fetching water, and better health due to improved availability of water for drinking and for personal hygiene (https://www.unicef.org/wash/malawi_102490.html).[2]

It is interesting to note that the World Bank, an international institution that provides loans to countries of the world for various capital projects, has contributed to providing more than 47 million people globally with access to an improved water source. Many of the countries that have benefited are developing nations such as Afghanistan, Tanzania, Croatia, Gaza *etc.* (http://www.worldbank.org/en/topic/water/overview#3).[3] Several non-profit and other NGOs have taken up the daunting task of ensuring that developing nations get access to clean drinking water. WaterAid is one such international non-profit organization that was set up in 1981 as a response to the UN International Drinking Water Supply and Sanitation Decade. This NGO has also contributed immensely with the help of its many local partners in providing clean and safe water to several developing regions such as the villages of Belavabary and Ambonidobo in Madagascar by building gravity fed water systems (https://www.wateraid.org/us/stories/clean-water-gets-the-party-started-in-madagascar).[4]

One noteworthy effort by the WHO has been the bringing together of various stakeholders, experts and practioners to produce simple and usable risk assessment and management tools that are applicable to small community water supplies. The is mainly because small community water supplies generally do not receive acceptable attention and that leads to the failing of water treatment and distribution systems. According to the WHO, over 2.5 billion rural dwellers are served by small water supplies and it could be a potential health hazard if they are managed by operators who lack adequate training and inadequate risk management practices. The efforts in this context by the WHO are based on the Water Safety Plan approach that is recommended in the WHO guidelines for Drinking Water Quality. Thus, initiatives like these by the WHO can not only lead to the prevention of illnesses and mortality and associated health costs but will also increase the possibility of livelihood activities, education and augment the long-term sustainability of water supply to small communities. Such an approach is very relevant for developing nations with populations mostly dwelling in a rural small community setting (http://www.who.int/water_sanitation_health/dwq/smallcommunity/en/index.html).[5]

This book is but a small attempt and the start on a voyage towards clean and safe drinking water for humankind. This compilation is an endeavor to gather and collate all possible information on available treatment methods particularly relevant to the developing countries. It is hoped that this book provides an insight on the currently available options of treating various physical, chemical and biological water pollutants along with their effectiveness and economics. With the aid of case studies that combine the crucial aspects of operation scale, cost and efficacy of various water treatment technologies, this book hopes to showcase the existing and potential possibilities of treating water, 'the elixir of life' for developing countries.

## References

1. http://www.unwater.org/publications/world-water-development-report-2018/.

2. https://www.unicef.org/wash/malawi_102490.html.
3. http://www.worldbank.org/en/topic/water/overview#3.
4. https://www.wateraid.org/us/stories/clean-water-gets-the-party-started-in-madagascar.
5. http://www.who.int/water_sanitation_health/dwq/smallcommunity/en/index.html.

# Subject Index